Innovations in Automotive Transmission Engineering

Martin G. Gabriel

Warrendale, Pa.

For permission and licensing requests, contact:

SAE Permissions
400 Commonwealth Drive
Warrendale, PA 15096-0001 USA
E-mail: permissions@sae.org
Tel: 724-772-4028
Fax: 724-772-4891

Global Mobility Database®
All SAE papers, standards, and selected books are abstracted and indexed in the Global Mobility Database.

For multiple print copies, contact:

SAE Customer Service
E-mail: CustomerService@sae.org
Tel: 877-606-7323 (inside USA and Canada)
            724-776-4970 (outside USA)
Fax: 724-776-1615

ISBN 0-7680-0995-2

Library of Congress Control Number: 2001099454

Copyright © 2004   SAE International

SAE Order No. T-109

Printed in the United States of America.

Other SAE titles of interest:

Electronic Transmission Controls
By Ronald K. Jurgen
(Order No. PT-79)

The Motor Vehicle Edition 13
By K. Newton, W. Steeds, and T.K. Garrett
(Order No. R-298)

Vehicle and Engine Technology
By Heinz Heisler
(Order No. R-233)

Automotive Handbook Edition 5
(Order No. BOSCH5)

Special Publication from the 2003 SAE World Congress:
Transmission & Driveline Systems Symposium 2003
(Order No. SP-1760)

For more information or to order a book, contact SAE at
400 Commonwealth Drive, Warrendale, PA 15096-0001;
phone (724) 776-4970; fax (724) 776-0790;
e-mail CustomerService@sae.org;
website http://store.sae.org.

Contents

Acknowledgments

I wish to express my thanks to my wonderful wife, Marie, for her support and encouragement in preparing this document. I am indebted to my daughter, Jeanne Heilman, for her precious time and inspiration in the proofreading. Many others have contributed, including Doug Cameron, Robert Cherrnay, Wayne Colonna, Don Fergle, Michael Imirie, V.J. Jandasek, Burt Martin, Walter Muench, Bruce Palamsky, R.C. Roethler, Karl Schneider, and Bruce Simpson. I wish to credit David Burhans, DLB Industrial Design, who created the excellent graphics. I appreciate the help of Martha Swiss of the SAE International staff.

Preface

An automotive transmission is broad ranging, strong but invisible within its own environment, the vehicle—like a tiger in its native haunts. For years, engineers have been developing transmissions, mated with engines, to produce the optimum vehicle powertrain. For all of their thousands of creative mechanisms and patented ideas, the transmission engineer's inventive genius has yet to meet all the needs of the customer and of society.

The automatic transmission has been in active development since the late 1930s. To commemorate the fiftieth anniversary of the automatic transmission, in March 1985, SAE International sponsored a panel discussion during its annual Congress held in Detroit, Michigan. Industry pioneers in the field discussed the latest transmission developments as well as the technical progress of the entire industry. And it is a major industry, currently measured in billions of dollars.

This book is designed to provide background and cohesive support for the manager who may have planning responsibility for directing the application of a transmission for a future vehicle application. Historical information is briefly reviewed as a basis for the state of development of current and future transmissions. Knowledge of past efforts helps to preclude repeating problems of the past. The format evolves naturally to the subject of transmissions of the future, identified as the "new" transmission. New transmission concepts are examined and tested to shed light on ways the engineer can help to meet the demands of the customer of tomorrow.

A transmission may take many forms and has various applications; therefore, for the purpose of the discussion in the following pages, some criteria have been established. First, the fundamental purpose of a transmission, as discussed herein, is to provide a cost-effective, efficient connection between a vehicle power source and its driving wheels. At first glance, this seems like a relatively straightforward objective. However, after vehicle criteria are established, some of which may appear to conflict among themselves, the problems to engineer a successful transmission, which will be accepted by the motoring public for a sufficient number of years to make it a profitable venture, may be quite formidable. The best example of conflicting criteria is to

economically produce a higher-performance vehicle engine–transmission with no compromise in fuel economy. As the following pages show, that ongoing objective is being met.

The discussion in the following pages will be limited to transmissions in passenger car or light truck applications, primarily in North America, to define a manageable scope of this vast and interesting subject.

Chapter One

Some History!

An early Mercury–Edsel–Lincoln technical service manual, in its description of a transmission, simply stated that a transmission is a device for adapting available engine power to road and load conditions. Far from a simple device, the transmission developed from a history of innovation that established its viability while remaining out of sight and "out of earshot" (see Appendix One).

Today, following years of development of automotive vehicles and their powertrains, we know that a transmission can be described by its role in many scenarios. Important characteristics, such as vehicle fuel economy, vehicle performance, and quality and reliability, to name a few, immediately come to mind. These and others will be identified and sorted out in the following chapters. Some characteristics affect the customer and driver of the vehicle directly. Others, such as fabrication complexity and cost, will directly affect the transmission engineer and manufacturer; however, the eventual purchaser of the vehicle is still affected, albeit indirectly. Therefore, how can management, and particularly the transmission engineer, favorably impact these issues before they become unmanageable or unacceptably costly?

In any event, over the years, after many transmission reiterations and billions of dollars invested, are we to find out that the early definition as previously stated is still the most accurate?

A review of the many transmission designs that have been implemented over the years reveals some interesting trends. Initial design approaches encompassed mechanisms to provide sufficient torque capacity to meet the torque output of the engine. Thus, the physical shape and overall dimensions of the transmission external case were defined. Then, with minor compromises, the vehicle underbody and connecting linkage would be adapted to accept that configuration.

In early designs, objectives such as performance and fuel economy were subrogated by customer convenience and smooth performance, compared with those of manual transmissions. Acceptable vehicle performance was attained by employing sufficiently

large engines with their correspondingly large displacements, as well as large drive gear ratios to ensure mobility over the common road grades. Maximum vehicle speed became a by-product of so-called engine redline limitations and overall drive ratios. Although the first transmissions developed torque multiplication deemed adequate for acceptable vehicle acceleration, whether by means of stepped gearing or belt or traction drive methods, the demand for driver convenience would define the parameters of the early automatic transmission. However, compromised vehicle performance or jerky shifting often was the result, discouraging full acceptance by the customer.

The marketing objectives of the vehicle and the overall vehicle powertrain arrangement dictate transmission configuration. A review of vehicle and transmission design history exemplifies certain objectives and compromises. For example, the well-known Ford Model T employed a front longitudinal engine location, with the transmission driving the axle and rear wheels by means of a longitudinal driveshaft. This rear-wheel-drive example of modular vehicle design featured the outstanding advantages of easily assembled components and dealer servicing. These features were appreciated by service technicians and home-garage repairmen alike, in comparison to the compact designs of today. Rear-wheel-drive vehicles performed their jobs well and are an economic success even today for various design and performance reasons.

A lengthy period of successful production rear-wheel-drive vehicles ensued, however with interruptions of new driving features, to be discussed later. Then, around 1959, the high-production British Mini introduced the transverse engine–transmission, driving the vehicle front wheels. The advantages of component integration, better controlled engine–transmission fabrication, vehicle packaging, and operating efficiency immediately became popular with the motoring public. Vehicle driving advantages, such as handling under snowy road conditions, ensured the popularity of front-wheel drive. Limitations such as greater vehicle weight distribution on the front wheels had to be addressed. New servicing techniques had to be developed.

It was natural that most early transmissions incorporated gearing to provide torque multiplication to accelerate the vehicle. The fixed nature of gearing has led to the incorporation of gear steps and an ancillary manual mechanism to enable the shifting from one gear ratio to another. It will be shown that alternative transmission mechanisms, such as belt and traction drives, can eliminate the gear steps and the associated potential jerkiness.

Geared transmissions have retained much of their acceptance since those early days. Ironically, it is not because of their simplicity but because of their durability and inherent efficiency. Correct gear-teeth design facilitates torque transfer by rolling tooth contact. Rolling action is practically frictionless, unlike transfer by friction belts or traction drives. And without significant friction, gearing provides the opportunity for excellent durability, thereby compensating for the inflexibility of the gearing. This will be discussed further in subsequent pages.

A review of developments over the years shows that more than 30 different transmission designs were produced in relatively high volumes, not counting heavy-truck, off-road, and specialty vehicles. Many of these are described in detail by P.M. Heldt [1-1], Phillip G. Gott [1-2], and others. Common features were as follows:

- A clutch or fluid member to connect the transmission to the engine

- A torque multiplication mechanism (usually a planetary geartrain)

- A device by which the driver controlled the transmission, engine, and vehicle

- A sealed case structure to house the transmission and ancillary components

These devices included shift mechanisms and hydraulic fluid systems for control, lubrication, and cooling. With these integral components—and with provisions for monitoring vehicle speed, engine power, and speed—was born the modern automatic transmission.

Of particular interest to us, however, are the significant new developments over the years and their impact on today's transmissions and the transmission design of the future.

A chronological look may begin with the significant impact of viable methods for shifting gears. Back in 1891, Emile Levassor was credited with a patent using external gears on parallel shafts progressively engaged and disengaged. Although this feature made the manual transmission more acceptable to the average motorist at that time, it was soon replaced by L. Renault's design that transmitted power in direct gear without going through an engaged gearset with its associated losses. This basic concept prevailed until another significant innovation, called the Synchromesh transmission, was produced by Cadillac in 1928. No longer would the driver have to deal

with the clashing of gears. The following pages will develop how electronics has facilitated shifting from gear to gear more smoothly, more reliably, and, most importantly, executing the shifts at the best time for maximum vehicle performance or for fuel economy.

Probably the next most significant development was the single plate clutch mechanism connecting the transmission to the engine. Other clutch designs came in the form of electromagnetic clutches, and multiple-plate dry and wet friction clutches. Subsequently, the fluid coupling appeared on the scene, first with the manual gearbox. With the advent of hydraulically augmented shifting and electronics, the clutch and gearbox drivetrain has become less popular. Thus was born the high-production fluid coupling automatic transmission such as the early General Motors' Hydramatic, with three simple gearsets, introduced around 1940. That gradually was superseded by various configurations of torque converter transmissions that contained four, five, or more individual elements or integral geared arrangements. Eventually, the three-element torque converter transmission won out, and it continues in its basic form to this day.

The three-speed torque converter automatic transmission endeared itself to the motoring public by providing a very acceptable torque multiplication range about equivalent to early three-speed manual transmissions and the four-speed coupling automatic transmission. During this period, when fuel economy was not an issue, the smooth-shifting automatic transmissions increased in popularity over the manual transmission, even though there were occasional reliability and maintenance issues.

References

1-1.   Heldt, P.M., *Torque Converters or Transmissions*, P.M. Heldt, Nyack, NY, 1942.

1-2.   Gott, P.G., *Changing Gears: The Development of the Automotive Transmission*, Society of Automotive Engineers, Warrendale, PA, 1991.

Chapter Two

The Transmission in Its Environment

In its role of adapting to the available engine power road and load conditions, the successful transmission must be designed with regard to the parameters established by the overall vehicle system.

Several engine–transmission–driveline configurations are available to the vehicle designer. The reasons behind the selection of these configurations are many: vehicle styling niche, vehicle utility, vehicle acceleration criteria, passenger seating, and luggage space. Adding to the difficulty of the selection process are the underlying fundamental considerations of manufacturing cost, quality and reliability, customer appeal, and servicing. Also, in today's global economy, vehicle weight and its impact on fuel economy must be considered.

To elaborate somewhat, a transmission must be properly configured to its vehicle utility niche. There are six generally accepted classes of automotive vehicles representative of more than 300 different global vehicles in production, and these are constantly evolving:

- Passenger cars—"A" thru "F" class

- Light trucks, such as pickup trucks and commercial vans

- Minivans

- Sport utility vehicles (SUVs)—typically four-wheel-drive (4WD) based on pickup chassis

- Crossover SUVs—based on car unibody chassis

- Medium- and heavy-duty commercial trucks

The range of these classes presents a challenge to the OEM in keeping the number of different production transmissions to a minimum. And yet, to adhere to vehicle marketing objectives and continue to maintain transmission overall performance and durability, proper matching is imperative.

After vehicle criteria are relatively defined, the transmission design can be established, whether it is to be a manual layshaft design, automatic, or a variation of the two. However, the current competitive environment demands additional selection criteria, such as the following:

- Vehicle styling niche—Powertrain architecture.

- Vehicle utility—Packaging of transmission of sufficient ratio range and torque capacity.

- Acceleration performance objectives—Impacts transmission ratio range and torque capacity. Each foot-pound of additional torque capacity contributes about a quarter-pound of weight, based on historical trends.

- Passenger seating and luggage capacity—Size of transmission housing and potential floor encroachment.

- Available manufacturing facilities—Possible transmission package constraints to accommodate existing dies for vehicle floor contour.

The underlying fundamental vehicle considerations of manufacturing cost, quality and reliability, passenger accommodation, customer appeal, serviceability, and fuel economy are basic, and their discussion in the context of transmission design is justified.

The so-called "conventional" vehicle engine–driveline in vogue in the early 1900s comprised a front engine location, suitable engine–flywheel connected transmission, and a rear driving axle connected by a longitudinal driveshaft. This conventional arrangement offers many advantages to the transmission designer and the vehicle designer. Its advantage of modular construction was recognized early, providing flexibility in component packaging, evenly distributed vehicle weight, and easier servicing by the mechanic.

Chapter Three

The Transmission: An Integral Part of the Vehicle Powertrain

Before discussing the conventional transmission arrangement further, what are some of the powertrain configurations available to the vehicle designer that may impact the transmission design? Discussed next are some of the most viable.

Rear-Wheel Drive (RWD)

Front Longitudinal Engine/Longitudinal Transmission Arrangement

As noted in Chapter Two, the front-engine rear-wheel drive (RWD) is the so-called conventional arrangement because of its successful history, recognition of its modular flexibility, and potential for balanced vehicle weight distribution. The transmission can be as long as required to contain the necessary gearing and clutches. Manual linkage to control the transmission functions could simply be of mechanical-type design. Field servicing is easy of access. The vehicle ride and handling are acceptable, except for some possible situations conducive to oversteer. However, this arrangement impedes optimum utilization of passenger and luggage compartment space. The long driveshaft connection to the rear axle not only encroaches into the floor, it presents potential noise, vibration, and harshness (NVH) issues. For years, these limitations were not given much shrift because, at least in the United States, the vehicle could be sized to meet occupancy comfort, with scarce regard for weight penalty. This penalty and the resulting compromise in meeting federal fuel economy requirements could be offset somewhat within the rules of corporate vehicle class averaging. Except for the large vehicles, the longevity of front-engine rear-wheel-drive arrangements continues to be threatened by various front-wheel-drive configurations, as noted later in this chapter.

Industry competition and customer demand for better vehicle ride and road handling are compelling vehicle designers to include independent rear suspensions instead of the conventional unsuspended axles. Cost comparisons on this basis have shown that,

in spite of its apparent simplicity, the rear-wheel-drive arrangement suffers a slight disadvantage to the front-wheel drive.[3-1]

Rear Longitudinal Engine–Transmission Arrangement

The rear engine–transmission orientation had been popularized in the Volkswagen Beetle and a few other small vehicles for its compactness. Its source of noise is well behind the driver. Issues include vehicle weight distribution and handling and restricted front luggage space. The rear orientation retains the features of better passenger space location and good rear-end collision protection.

Front-Wheel Drive (FWD)

Front Longitudinal Engine/Longitudinal Transmission Arrangement

In a front-wheel-drive (FWD) longitudinal engine vehicle, there are essentially two practical transmission configurations: (1) an in-line engine/transmission layout, similar to the conventional as discussed previously, and (2) a U-drive architecture.

The longitudinal arrangement is attractive for its adaptability to a transfer drive for an all-wheel-drive (AWD) architecture. Packaging space available to the in-line transmission can be limited by the driver/passenger floor, if not the differential. The differential is accommodated, either within the gearbox or ahead of it next to the engine, with appropriate length half shafts and constant velocity (CV) joints.

The second configuration comprises a chain-driven transmission in parallel with the engine in a U-drive arrangement. It was popular because it could accommodate more transmission gear steps with a larger displacement engine. Package complexity and potential handling issues due to unequal half shafts may have led to the eventual demise of this configuration.

Transverse Engine/Transverse Transmission Arrangement

This configuration gained its popularity with the small displacement engine vehicles because of its compatibility with both automatic and manual-type transmissions and the continuously variable transmission (CVT), as well as for its contribution to vehicle crashworthiness. The engine and transmission are generally placed ahead of the front half shafts. The limited track space available between the front wheels has led to creative compacting of the transmission components along its axis. This has cleared the way for applications in higher-engine-power vehicles.

The hybrid engine simple planetary split torque configuration with double motor/generators, as in the Toyota Prius vehicle, lends itself to this transverse package configuration. Satisfactory weight distribution is realized by locating the heavy battery at the rear. Another parallel configuration augments engine torque with a Simpson gearset plus a single motor/generator with potential for four regeneration modes, as described by N. Higuchi [3-2] of Honda.

Four-Wheel Drive (4WD)

The four-wheel-drive (4WD) on-demand vehicle or all-wheel-drive (AWD) in constant engagement configuration is generally associated with large sport utility vehicles (SUVs), although its popularity in crossover vehicles is growing. The engine is located in front and mounted either longitudinally or transversely with the transmission in-line or transverse. Although not necessarily integrated with the transmission case, the transfer drive may be connected to the shafts by means of a wet-plate disk clutch, electromagnetic clutch, or viscous coupling. Without getting into the details of all the possible arrangements, the grand objective is to provide sufficient drive torque to the individual wheels in response to both on-road and off-road driving conditions and handling situations. Assigned to the electronic control unit (ECU), the objectives are met by channeling vehicle throttle, vehicle velocity, acceleration, and so forth through appropriate algorithms to distribute optimum torque to each wheel. Inherited from rear-wheel-drive (RWD) systems, many vehicles, especially small SUVs, use systems based on transverse front-wheel-drive chassis.

The impact of full-time 4WD on fuel consumption due to the additional weight of the transfer drive system seems to be greatly mitigated by the owner's satisfaction in vehicle usefulness. An electronic 4WD on-demand option augmenting operation in conventional two-wheel drive (2WD) is gaining acceptance. This precludes the efficiency loss of internal torque loading that can occur due to differences in wheel speed rotation in situations such as vehicle turns. One application of 4WD on-demand provides for an electromagnetically controlled clutch system at the rear axle.[3-3] A special control algorithm applies the clutch to drive the rear wheels to augment the normal front-wheel drive when wheel slip is indicated or cornering yaw requires it. With this configuration, drive performance is less likely to be affected by changes in

weight distribution due to passenger occupancy. The overall package is about 30 kg (66 lb) lighter than a full-time system, resulting in fuel savings as well.

Transmissions used in four-wheel-drive applications generally require stronger components and more firm calibration compatible with the more severe usage of which the SUV is capable. Transmission controls can be tailored to meet SUV performance requirements, particularly if a two-speed transfer case is employed.

The increasing acceptance of production hybrid powertrain vehicles paves the way for locating an additional electric motor in the rear to drive the rear wheels, thus expanding the hybrid application to an effective 4WD minivan vehicle.[3-4] A significant advantage follows with the elimination of the need for a longitudinal driveshaft and floorboard hump usually required with this type of vehicle.

References

3-1.   Seznec, H., and LaGrange, H., "The Technique of Front-Wheel-Drive in Europe," SAE Paper No. 750013, Society of Automotive Engineers, Warrendale, PA, 1975.

3-2.   Higuchi, N., et al., "Development of a Novel Parallel Hybrid Transmission," SAE Paper No. 2001-01-0875, Society of Automotive Engineers, Warrendale, PA, 2001.

3-3.   Ohba, M., et al., "Development of a New Electronically Controlled 4WD System...," SAE Paper No. 1999-01-0744, Society of Automotive Engineers, Warrendale, PA, 1999.

3-4.   Kondo, K., et al., "Development of an Electrical 4WD System for Hybrid Vehicles," SAE Paper No. 2002-01-1043, Society of Automotive Engineers, Warrendale, PA, 2002.

Chapter Four

Transmission Types

As stated previously, there are at least six classes of automotive vehicles, which require appropriate powertrains with their respective engine and transmission combinations. To conform to these various vehicle applications and meet customer needs, several basic transmission types have stood the test of time.

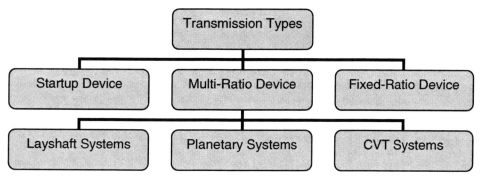

Figure 1. Transmission classification chart.

As illustrated here in the transmission classification chart, a complete transmission comprises a startup device, a multi-ratio device, and a fixed-ratio device. The startup device may be a friction clutch or electromagnetic clutch, a hydrodynamic device such as a torque converter, or an electric motor. The fixed-ratio device may be a rear axle or differential. The chart classifies transmissions first according to their basic functions and then according to the type of torque multiplication mechanism. These may be categorized even further as to whether they are torque break systems or non-torque break (i.e., powershift) systems.

Some multi-ratio devices or transmission types and their basic components are briefly reviewed next as a foundation for projecting new directions and innovations.

1. Manual-step layshaft transmission

2. Automated shift transmission (AST)

3. Clutch/flywheel startup device

4. Fluid coupling transmission

5. Torque converter transmission

6. Continuously variable transmission (CVT) with belt or chain variator

7. Continuously variable transmission (CVT) with toroidal variator

8. Continuously variable transmission (CVT)—hydrostatic

9. The hybrid

10. Power-split configuration

Manual-Step Layshaft Transmission

The task of harnessing the ubiquitous gasoline engine in the automobile has been addressed in many ways. It naturally fell first to the aptly called manual transmission. The development of the manual transmission is well documented and is still continuing. For example, the Emile Levassor early manual transmission was patented in 1891.[4-1] A basic offering for many years, because of its versatility and reliability, enhancements to the manual transmission continue to be made. Importantly, the comparatively plain manual transmission possesses some useful features that continue to be the wellspring for the transmissions of tomorrow.

Latent advantages of the manual transmission that have not escaped notice are its modular design adaptability to a range of gear ratios and its excellent power efficiency. Torque multiplication is accomplished by tooth-to-tooth near-rolling contact with well-designed gearing. Rolling contact has the potential to provide the most efficient power transfer compared with partial sliding contact. This is a distinct advantage over roller-and-pulley arrangements that rely on friction contact means for torque transfer and multiplication, such as in cone-type friction drives or chain drives. This is discussed later in this chapter.

Historically, the ratio range of the manual transmission has been achieved by means of suitable gearing, mounted on parallel shafts. The design has almost unlimited flexibility and range capability. Changes from one gear ratio to the next generally are accomplished manually. Very early manual transmissions required driver proficiency to shift from gear to gear by engaging dog clutches. Then, in the early 1930s, synchronizers were introduced, facilitating relatively smooth engagements even by the inexperienced driver.

Supplementary aids may be added to the basic manual transmission to assist the driver by reducing shift efforts and minimizing shift feel. Through evolution of design, features have been added to execute the transmission shift events at the best time for every performance demand and every vehicle speed. The automatic transmission is the eventual result. In addition to providing a pleasing drive to the customer, vehicle fuel economy becomes an unforeseeable, important by-product of the automatic transmission.

Automated Shift Transmission (AST)

With the growth of electronics and the introduction of electronic engine control and subsequent electronic transmission control, the automated shift transmission (AST) is becoming a viable alternative to the conventional manual transmission, if not the automatic planetary type transmission.[4-2] It has the additional attraction of being able to utilize existing manual transmission manufacturing facilities. The AST is a significant step beyond the automated manual transmission (AMT) because torque interruption is reduced or completely eliminated during the shifting sequence, enhancing both driver and passenger satisfaction. As the name implies, the transmission shifts are performed with electronic control unit (ECU) driven actuators communicating with the engine electronic throttle control unit in a semi-automatic manner, still retaining some driver shift control.

The ideal AST has the potential to provide the following features desired by the driver:

- Efficiency and vehicle fuel economy usually attributed to the conventional manual transmission

- Ratio range and vehicle performance, limited only by the number of gear steps

- Almost jerk-free vehicle shifts

A dual clutch transmission (DCT) system to engage and disengage parallel layshaft gearing is becoming the architecture of choice, yielding the ultimate potential of an automated shift mechanism (ASM) with innovative electronic shift control synchronized with engine throttle control. The components may include the following:

- Addition of a second dry clutch to control the second layshaft

- Double wet clutch system, as a viable alternative to the dry clutch system

- Electromechanical clutch system

- Electric control motor for activating axial gear-to-gear engagement

- Any combination of the preceding four components

The double clutch layshaft transmission system has its followers, as it inherently avoids the open clutch drag losses associated with a multi-step planetary configuration. It has the attribute for precise control, particularly where a limited slip is indicated for engine noise, vibration, and harshness (NVH) control. It also has the potential for less wear over the life of the layshaft gearing.

The AST is particularly attractive on the European scene because of its adaptability to the local production manual-transmission facilities, but it is not without some compromises. An elaborate electronic controls system is required to integrate engine, vehicle, and driver feedback to perform the shifting from one layshaft gearset to the other, adding significant complexity. The configuration retains the high efficiency and flexibility for multiple speeds of the basic manual transmission.

Table 1 summarizes the attributes of some of the current types of transmissions. Viability of the major internal components, such as the fluid pressure pump, clutches, gears, and controls, will be discussed in subsequent chapters.

Clutch/Flywheel Startup Device

One of the most significant early developments was the startup device connecting the transmission to the engine. The challenge was to make the connection smoothly, disengage when required, and then repeat the cycle. A myriad of shifts are demanded by a driver using a manual transmission, such as fast power upshifts, skipped-gear shifts, and forward-to-reverse vehicle rocking out of snow. The spring-loaded dry friction clutch, actuated by means of a foot-operated linkage, was one of the first types of mechanisms called upon to do this task.

TABLE 1
CURRENT TYPES OF TRANSMISSIONS

Transmission	No. of Ratios	Ratio Span	Efficiency	Fuel Economy*	Cost*
Manual Layshaft	3 to 6+	2.5 to 6+	95%+	Base	Base
Automated Manual Transmission (AMT)	Equal	Equal or +	Equal	Base+	Base or +
Automated Shift Transmission (AST)	6+	5+	90–95%	Base or +	Base++
Conventional Planetary Automatic Transmission (A/T)	2 to 7	2 to 6+	60–90%	Base– to base	Base++
Continuously Variable Transmission (CVT)	Infinite	5+	80–95%	Base to base+	Base+++

* Based on Table 1 of Reference 4-2.

This is an opportune place to discuss the role of the clutch, as well as the flywheel, in the startup role of transmitting torque from the engine to the transmission. This role recently has been receiving much more attention. The clutch/flywheel essentially has two functions: (1) provide a means for smooth engagement and disengagement of the layshaft transmission or variator with respect to the engine, and (2) provide sufficient inertia mass to smooth and maintain engine rotation during engine start and idle.

To provide for the engagement and disengagement of the transmission from the engine, the conventional method has been to utilize a single or double plate clutch smoothly applied by a highly developed spring and disengaged by means of appropriate manual linkage. This method is attractive for its inherent capability for complete disengagement with essentially no drag. Elimination of drag can be attractive with new vehicle systems that automatically disconnect the engine when it is turned off at a vehicle stop for fuel economy reasons.

The function of engaging the layshaft gearbox to the engine generally is associated with a conventional single or double plate-type dry clutch, but it is not limited to the dry clutch. In fact, the multi-plate wet clutch, which has been so successful in automatic transmissions, has some useful characteristics for this function. Design torque capacity of multi-plate clutches may be increased simply by providing additional plates. The smoothness of engagement may be precisely controlled. Satisfactory

clutch durability may be attained with good plate design, good fluid friction characteristics, and provision for cooling. Packaging limitations and controls complexity of twin clutch applications must be considered in planning the entire vehicle system.

The second role, providing startup inertia mass, is dependent on the characteristics of the engine. Although the need could be the topic for another book, controlling flywheel inertia cannot be underestimated in view of its effect on vehicle acceleration and fuel economy.

Other engine-to-transmission clutch systems are gaining interest as new applications become apparent. The electromagnetic-type starter motor can harness vehicle mass to power an electric generator during the braking function to avoid losing that energy to drive the engine—an energy conservation technique. This type is particularly attractive for its torque capacity and potential to augment vehicle startup in a semi-hybrid arrangement. The weight and inertia of an electromagnetic system is a disadvantage, as is the cost.

Although the hydrodynamic startup device such as a fluid coupling is being discussed subsequently, it is worth mentioning here as an option for connecting the gearbox to the engine instead of the conventional friction clutch. This option takes advantage of the inherent capability of a fluid coupling (or a torque converter) to dampen engine torsional vibrations and has been used in series with various multiple gear ratio devices. However, the subsequent development and refinement of the torque converter, with its range of torque multiplication, has in effect superseded the fluid coupling.

Fluid Coupling Transmission

One of the early mass-production automotive applications of the fluid coupling was the 1939 Hydramatic transmission. It was the first fully automatic passenger car transmission, and its application was expanded to millions of other vehicles. Later, the 1956 Hydramatic incorporated a second, smaller dump-and-fill fluid coupling for smoother 1–2 shifts that involve a greater gear step. Fluid couplings provided the advantages of smooth vehicle startup, capability to idle the engine at traffic stops while in gear, damping of the engine torque impulses, and smoothing of shifts from gear to gear. Although it was recognized that the fluid coupling has some inherent disadvantages, such as slip under load with no torque multiplication, the disadvantages were partially offset in the design by providing an additional lower transmission gear ratio and by splitting the engine torque between the coupling and the gearbox.

The fundamental relationship for a hydrodynamic unit is

$$T = CN^2D^5$$

where

T	=	engine torque
C	=	a constant
N	=	engine speed
D	=	coupling nominal diameter

Therefore, as shown here, for a given fluid coupling size or diameter D, coupling efficiency is approximately inversely proportional to the input torque. This is the incentive for splitting the torque between the coupling and mechanical drive

$$\text{Efficiency} \approx N^2/T$$

Since the early Hydramatic, there have been several transmission arrangements using both the coupling and torque converter, which have utilized the split torque concept.

The fundamental two-member fluid coupling has lost favor in transmission applications, primarily because of the diameter required to carry modern levels of engine torque efficiently. In addition to its ability to dampen engine torsional impulses, the usefulness of hydrodynamic unit characteristics has been broadened with various different blade configurations, with variable fluid fill and by adding components such as a stator—thus creating a torque converter. A version of the fluid coupling has been employed effectively as a downhill braking retarder in heavy-truck transmission applications.

Torque Converter Transmission

The hydrodynamic torque converter has a long history of successful applications by the military, the commercial vehicle industry, and eventually in conventional passenger car vehicles.[4-3] It has the capability to multiply torque, fluid-smooth within its design range. Torque multiplication is greatest at stall condition or vehicle startup and decreases with output speed to 1:1. At this point, fluid coupling range is attained by employing a one-way clutch that permits the stator to free-wheel out of the way. Special transmission architectures are possible by gearing the torque converter elements in various split torque relationships, similar to a CVT. However, these have

had limited application because of redundancy and the limited hydrodynamic operating range of fixed, bladed elements.

Similar to the fluid coupling, the torque converter has the capability to dampen the torsional impulses created by the internal combustion engine. Torsional damping is particularly necessary to reduce noise, vibration, and harshness (NVH) with modern, highly developed internal combustion gasoline and diesel engines. Torsional impulses of a high-compression ratio engine, such as a six-cylinder engine, take place three times for every revolution. The impulses become more acute as power-to-weight ratios increase and must be reckoned with, especially in diesel engine drives.

Although not always acknowledged, the basic torque converter has excellent durability compared to a wet or dry clutch startup device, primarily due to the fluid environment in which it operates.

Stall torque multiplication for automotive three-element production torque converters generally has been in the range of 1.8:1 to 2.4:1. However, stall torque multiplication can be increased well beyond that by adding integral stages of turbine and stator components, albeit with some compromise in torque converter efficiency. In fact, an early mass-produced torque converter transmission employed no supplementary gearing in the vehicle normal drive range.

In addition to stall torque multiplication, the overall performance characteristics of a torque converter are greatly determined by the shape and profile of the bladed elements. Much analysis and development is necessary to create optimum blading within the limitations created by the capability to be mass produced. Pioneering that effort, I engineered and placed into operation at Ford Motor Company in 1962 a novel full-size hydraulic flow tunnel apparatus to test torque converter blading, as shown in Figure 2(a).

As in the fluid coupling, a useful characteristic of the torque converter is the shape and value of the K or capacity factor, broadly defined as the input speed divided by the square root of the input torque. When directly connected to the engine in the vehicle, the value of this characteristic establishes the so-called stall speed of the engine within the full-throttle engine speed profile. In direct gear, the K factor also defines the degree of slip between the input and output shafts, thus impacting transmission efficiency and vehicle fuel economy.

Figure 2(a). Hydraulic flow tunnel blade test apparatus.

Figure 2(b) illustrates the K factor, with torque ratio and efficiency plotted with respect to speed ratio for various three-element torque converters. Of particular significance is the difference in speed ratio at a selected high K value corresponding to a vehicle road load condition. Reading from the figure, for a torque converter K equal to 220, which approximates vehicle road load power requirement at 30 mph (assuming direct gear), the transmission slip can range from 7% (0.93 speed ratio) to 16% (0.84 speed ratio). This is more than a 100% increase in slip, contributing to a significant increase in road load power loss and vehicle fuel economy, and it corresponds to approximately 6% loss in metro-highway fuel economy.

To reduce torque converter losses due to slip, virtually all transmissions since the 1980s have a built-in electronically controlled lock-up clutch. In earlier installations, lock-up was applied only in high gear; however, current practice generally applies the lock-up clutch in all gears except the first. The lock-up clutch assembly generally consists of a hydraulically applied single- or dual-plate disk clutch built integrally within the torque converter, so that the turbine turns at engine speed. The plate assembly includes a tuned spring damper to isolate torsional vibration in lock-up range. The resulting effect is similar to a wet clutch engaging a layshaft transmission. Clutch engagement is electronically scheduled to take effect under accelerating conditions usually in coupling range when torque multiplication is no longer available. The clutch is disengaged under torque demand situations or at coast-down engine speeds at about 900 rpm, thus avoiding the possibility of annoying engine harshness that can occur at low speeds under load.

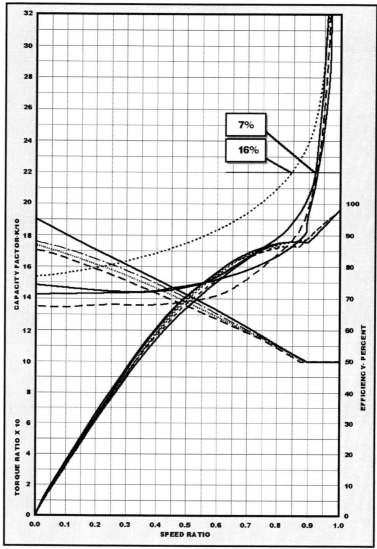

Figure 2(b). Torque converter K shape effect on slip.

Although the lock-up clutch feature is not new, it has been plagued with problems of so-called "shudder," excessive slip, friction facing wear, and noticeable engagement-disengagement feel. Development over the years has reduced these problems significantly. As a fix for the shudder problem in the field, transmission rebuilders generally replace the transmission fluid with new fluid of the original friction characteristics. This can become a maintenance issue. Resolving more severe concerns can involve cutting open the torque converter, replacing the clutch components, and rewelding the torque converter while tearing down the entire transmission to clean out the debris—a major job. Of course, the customer usually is given a replacement transmission to get him on his way.

An alternative design solution would be to reconfigure the overall planetary transmission architecture to relocate the lock-up clutch, of larger capacity, inside the gearbox area and reverting back to the basic converter configuration. This also would permit reducing the torque converter diameter in accordance with current performance trends.

Since the late 1970s, the role of the torque converter clutch has been expanded to perform a second function: controlled limited slip under load. This slip, electronically controlled to limit temperature buildup, improves torsional damping and improves fuel economy by enabling lower engine speeds. The ideal lock-up electronic control contributes to optimum overall vehicle performance, particularly vehicle fuel economy, without negating the features of the torque converter and without significant engagement and disengagement feel.

An alternate design to contain torque converter slip at high speeds utilizes a viscous shear damper, in an integrally sealed chamber within the torque converter, filled with special silicone fluid. This eliminates the problem of wear but provides for no external control.

Continuously Variable Transmission (CVT) with Belt or Chain Variator

There are several ratio-changing transmission concepts that depend primarily on the friction between components to execute the load-torque transfer. Historically, fixed-center, fabricated V-belt/pulley arrangements have long been employed in industrial applications and light agricultural vehicles. Power is transferred from one adjustable sheave or pulley to a second adjustable pulley by means of friction between the belt and the pulleys. Torque multiplication is accomplished by concurrently decreasing the axial distance between the faces of the drive pulley and increasing the driven pulley face distance, thereby varying the radius at the belt–pulley contact. Speed increase is accomplished by reversing this sequence.

With greater durability and capacity for automotive applications, the push belt type variator drive has become a successful means for providing the CVT function. Known as the Van Doorne design, it employs multiple, radial steel segments operating in compression and retained against the pulley face by two pairs of up to twelve laminated precision steel bands.

Figure 3 illustrates the basic architecture of a push belt CVT application.[4-4] The push belt variator provides good durability and torque capacity through better distribution of belt stresses compared with the original one-piece composite belt designs. Operating stress of the precision bands is a primary factor that determines belt torque capacity.[4-5] The design torque criterion is the stress of the innermost band at the driving pulley operating at the lowest speed ratio and determines the critical belt-arc of engagement.

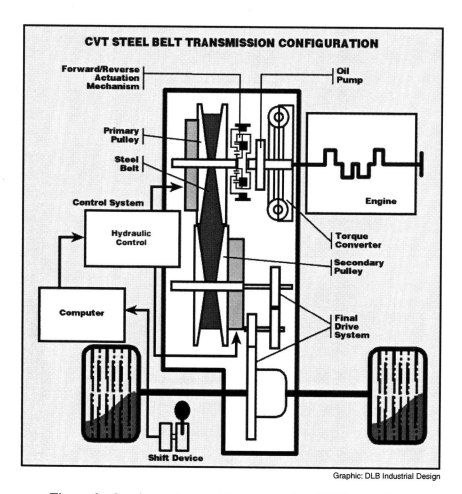

Graphic: DLB Industrial Design

Figure 3. Continuously variable transmission (CVT) belt drive.
(Source: Reference 4-4, Figure 2.)

For this belt traction drive, or any CVT, to provide optimum fuel economy and minimum engine emissions requires the precise control of engine fuel, matching drive ratio and engine throttle demand. This applies to the entire vehicle operating range from road load to full throttle acceleration, as well as during deceleration. An ideal controls system for a pleasant customer driving experience takes the inertia of all the components into account, including the engine rotating components, transmission pulleys, and the driveline. It has been shown that inertia and ratio change lag can essentially be offset with an integrated engine controlled torque burst.[4-6] For optimum operation of this drive, the controls are designed so that the engine runs at near-constant speed, matching throttle position at near minimum specific fuel consumption levels, achieving the best fuel economy and overall performance.

The traction drive has the flexibility to achieve overall ratios of about 6:1 in practice. As a result, metro-highway fuel economy gains of 10% have been reported compared to a four-speed automatic transmission in a two-liter engine installation.[4-7] Fuel economy levels were comparable to those using a five-speed manual transmission.

The mechanism of torque transfer between the driving pulley and the belt and subsequently to the driven pulley is primarily dependent on the torque load, friction coefficients of the segments, band, pulleys, and the pulley clamping force. This clamping force establishes the operating radius of the belt–pulley interface and must be adequate to ensure that torque transfer occurs without harmful belt–pulley slip. On the other hand, loss of efficiency due to excessive clamping force may result by attempting to reduce slip below feasible levels (approximately 3–4%) at maximum operating torques and ratios.

The LuK chain drive is a viable alternative to the push belt. It utilizes a specially fabricated chain running in tension over adjustable pulleys. Good torque capacity-to-mass ratio is possible because of its rolling-pin pulley contact principle. However, durability may be more sensitive to "slip-shock" than the belt design because of the limited pin-to-pulley contact. The usual NVH considerations associated with a chain drive exist due to the pitch of the chain links.

Both the belt and chain drive variators depend on adequate, directed lubrication and accurate control of clamping. To attain the maximum potential of either drive requires matching torque demand with engine speed and engine fuel island characteristics according to vehicle performance and fuel economy objectives.

Although CVT technology offers simplicity and eliminates the issues associated with gear shifting, reservations for its universal acceptance have been expressed. Bob Welding, president of Borg Warner Transmission Systems, notes [4-8] that fuel economy gains over the conventional automatic transmission have not been as significant as projected, only 3–4% instead of his anticipated 7–8%, and fabricating costs have been higher than projected because of the precise manufacturing tolerances involved.

Continuously Variable Transmission (CVT) with Toroidal Variator

The toroidal CVT is a common type of traction drive comprising an input disk or cone and an output disk, both in rolling contact with adjustable power rollers. Drive is transmitted by means of the effective tangential forces developed at the rolling contact areas between the rollers and the cones. Torque multiplication is varied by altering the angle of contact, and thereby the operating radius between the power rollers and the cone. The natural longitudinal architecture of the disk assemblies makes the toroidal transmission particularly suitable for a front-engine, rear-wheel-drive vehicle application.

Within limits, the roller contact loads may be increased to augment the tangential force and thereby the torque capacity of the toroidal drive. One limitation is the support structure and additional pulley mass required to minimize stresses and deflections.

Hertzian stresses of up to 4 GPa may be warranted at the contact between the rollers and cavities, based on ball-bearing experience. Similarly, the surface finish and metallurgy must be of extremely high quality. Torque transfer in a traction drive is proportional to the number of rollers, the contact radius, the friction coefficient, and the normal force at contact

$$T = CN_r R_c \mu F_n$$

Because three adjustable rollers is the maximum number that can practically be accommodated and the friction coefficient is practically limited to about 0.1 and the normal force is limited by the allowable contact Hertz stress, the limiting torque for a given ratio span is essentially proportional to the maximum roller-cavity contact radius

$$T \approx R_c$$

Accordingly, the torque capacity is limited by the size of the toroidal variator within the space available in passenger car applications, unless the power flow is split.

To operate satisfactorily at the high tangential driving forces and contact stresses that are present, special elastohydrodynamic fluids with effective friction characteristics up to 50% higher than conventional automatic transmission fluid (ATF) are necessary. Note that this pressure-lubricated rolling contact between the rollers and cavities offers the advantage of quieter ratio change compared to shifting gears.

The success to date of the toroidal CVT has been achieved by resolving the intricacies of the ratio-changing mechanism and the related electronic controls to match ratio to engine and driver demands.

Internal axial loads and moments are kept in balance by utilizing opposing input and output power rollers. Due to internal space limitations, the number of rollers per set of cavities is limited to three. Each roller must be precisely located and angularly controlled with respect to the others to preclude durability issues caused by perturbations from rolling contact at infinitesimally different radii. This can be a significant design challenge, with the metallurgy and lubrication of the components in stressed contact.

Although not a part of the toroidal variator, some provision to ensure smooth vehicle startup and prevent driveline shock may be justified because of the inherent slip constraints of the contact between the rollers and cones. For example, the Nissan-JATCO transmission has incorporated a torque converter connecting the engine and the half-toroidal variator for smooth startup and torsional engine damping.[4-9] To help isolate driveline shock, adding a shock absorbing tube at the connection to the power roller trunnions was found to be effective.

As opposed to a half-toroidal variator geometry, the full-toroidal mechanism has the flexibility for a wider speed ratio range from forward through neutral to reverse drive, resulting in greater CVT versatility. A design by Torotrak incorporates a split-torque feature that allows forward, reverse, and geared neutral without a vehicle startup device such as a torque converter.[4-10] However, the increased range of contact speeds and loads and roller angular control create unique design issues. These issues are less formidable in the half-toroidal variator design because of the smaller range of roller angle required.

Continuously Variable Transmission (CVT)—Hydrostatic

The hydrostatic transmission has been studied in earnest for automotive applications almost since the beginning of the industry. It basically consists of an engine-driven, positive-displacement pump that transmits power via fluid under high pressure to a positive-displacement motor to provide a CVT.

Hydrostatic transmission arrangements are attractive for their capability to provide broad, variable torque multiplication range by controlling the displacement of the pump or both units. They offer flexibility of location limited only by the pressure lines.

Various types of pump/motor arrangements have proven to be feasible. Designs with radial cylindrical pistons, ball pistons, axial, and bent-axis piston configurations with adjustable swash plates, as well as adjustable vane types, have been developed with various degrees of success. Operating pressure levels up to 7000 psi (approximately 48,000 KPa) may be employed to attain acceptable operating efficiencies, on the order of 90%. These high pressures present significant design and operating issues. For example, to contain the high pressures requires components of weighty construction to minimize distortions. Unless hydraulic passages are well refined to minimize cavitation, noise levels due to the turbulent fluid flow between the pump and motor are difficult to suppress.

Of the various hydrostatic design types, the piston/swash plate pump/motor, with its greater potential for acceptable volumetric efficiency, continues to dominate interest. The bent-axis type is attractive for its inherent potential for lower piston side loading. To offset unavoidable hydraulic and mechanical losses, particularly at high relative speeds with high internal fluid velocities, split torque designs in which some power mechanically bypasses the hydrostatic unit have shown to be attractive.

The jury is still out as to whether a robust mechanism such as a hydrostatic transmission installed in passenger vehicles that are always under significant power-to-weight constraints to meet fuel economy targets can be a marketing success. This is especially true if the hydrostatic transmission must operate in a regenerative mode part of the time, resulting in a degree of redundancy. As noted, issues such as hydraulic noise levels caused by internal pressure gradients, lower road load efficiencies, and relatively high mass of the system have shown to be difficult to resolve for automobile drives, compared with agricultural applications.

The Hybrid

The hybrid powertrain has crept up on the scene so persistently that its impact is not going unnoticed. Although, to date, the drive basically comprises a low-power engine/electric motor–battery combination, the architecture is proving itself in production in various vehicles. Management of the interacting drive components via exhaustive electronic control technology has been shown to be the key.

At least three types of hybrid configurations employ a conventional engine in combination with a permanent magnet synchronous motor-generator and electric inverter:

1. A series type, with both the engine and electric motor driving the wheels in series

2. A parallel type, with the engine augmented by a motor-generator capable of driving the wheels

3. A power-split type, with engine power divided between driving the wheels and charging a battery via a generator to provide power to a motor that concurrently drives the wheels according to optimum power management control.

Variations of these three configurations have been proposed for specific applications, such as the capture of regenerated power for improving fuel economy. An example of this is the Eaton HLA system for light- and medium-duty applications utilizing low- and high-pressure accumulators in a parallel connection.[4-11] Composite fabricated material is used for the accumulator shells to help offset the weight penalty.

Power-Split Configuration

First produced in 1997, the Toyota Prius front-wheel-drive configuration is well known. Utilizing a power-splitting planetary gear behind a four-cylinder 1.5-liter engine, the synchronous motor-generator provides vehicle startup, after which the engine essentially takes over under various alternate drive and battery charge situations with the aid of an appropriate inverter. Approximately 30% of the energy normally lost in decelerating the vehicle is recaptured and subsequently used to power the motor. To coordinate the functions of drive, power regeneration, and optimized utilization of energy requires a coordinated electronic control of inputs of the engine, motor, generator, battery, vehicle speed, vehicle acceleration, and so forth. Judging from sales trends, vehicle performance meets the expectations of most purchasers. It

is only a matter of time before sales of vehicles of this type of architecture increase with larger engine displacements and with more efficient electrical power storage and transfer. The favorable public response is an indication of the acceptability of these hybrid drives utilizing appropriate control algorithms to accomplish the shift events between the engine and motor and back.

Another application of a power-split configuration is in hydraulic pump/motor drives as practiced in agricultural tractor applications and recently described for a Honda Accord vehicle application. Its feature of infinite variability obviates the need for a hydrodynamic startup device.

References

4-1.　Gott, P.G., *Changing Gears: The Development of the Automotive Transmission*, p. 8, Society of Automotive Engineers, Warrendale, PA, 1991.

4-2.　Link, M., et al., "The Automated Shift Transmission (AST)—Possibilities and Limits in Production-Type Vehicles," SAE Paper No. 2001-01-0881, Society of Automotive Engineers, Warrendale, PA, 2001.

4-3.　Jandasek, V.J., "Design of Single Stage, Three-Element Torque Converter," SAE Seventh L. Ray Buckendale Lecture, Society of Automotive Engineers, Warrendale, PA, January 1961.

4-4.　Abo, K., et al., "Development of a Metal Belt-Drive CVT Incorporating a Torque Converter for Use with 2-Liter Class Engines," SAE Paper No. 980823, Society of Automotive Engineers, Warrendale, PA, 1998.

4-5.　van der Sluice, F., "Stress Reduction in Push Belt Rings Using Residual Stresses," SAE Emerging Technologies TOPTEC Symposium, August 12, 2003, Society of Automotive Engineers, Warrendale, PA, 2003.

4-6.　Yasuoka, M., et al., "An Integrated Control Algorithm for an SI Engine and a CVT," SAE Paper No. 1999-01-0752, Society of Automotive Engineers, Warrendale, PA, 1999.

4-7.   Boos, M., and Mozer, H., "ECOTRONIC—The Continuously Variable ZF Transmission (CVT)," SAE Paper No. 970685, Society of Automotive Engineers, Warrendale, PA 1997.

4-8.   Comments by Bob Welding, president of Borg Warner Transmission Systems, *AEI*, May 2002, p. 46.

4-9.   Nakano, M., et al., "Development of a Large Torque Capacity Half-Toroidal CVT," SAE Paper No. 2000-01-0825, Society of Automotive Engineers, Warrendale, PA, 2000.

4-10.   Fuchs, R.D., "Full Toroidal IVT Variator Dynamics," SAE Paper No. 2002-01-0586, Society of Automotive Engineers, Warrendale, PA, 2002.

4-11.   Lyman, Richard, and Bohlmann, B., "Hybrid Hydraulic Powertrains: History, Current Status, and a Vision of the Future," SAE TOPTEC, August 13, 2003, Society of Automotive Engineers, Warrendale, PA, 2003.

Chapter Five

Gearing the Transmission

The elemental multiple-tooth gear has long established itself as an efficient means for torque transfer. It naturally became an integral part of most transmissions, manual or automatic. From parallel shaft manual gearboxes, it was a natural evolution to the planetary gearset with its inherent compact architecture as used in many automatic transmissions.

Exactly what is the optimum gearing arrangement for a transmission? Is there only one answer to this question? Perhaps not. If the relevant criteria are sorted out and weighed according to the transmission–vehicle application, the optimum gearing arrangement can come into focus. The basic transmission arrangement, whether manual or automatic, front-wheel drive or rear-wheel drive, generally comprises a series of forward speeds and a reverse gear. The forward gears usually include an overdrive ratio.

It is significant to note that most recent automatic transmissions include an overdrive ratio. Other than fuel economy, what is the rationale for this trend? It is possible to design gear arrangements with overall gear ratio spans of 4:1 or greater, equivalent to an overdrive arrangement. There are two not-so-obvious reasons: (1) Increasing the transmission lower gear ratios to match the ratio span of an overdrive arrangement adds component bulk to handle the added torque, and (2) The rear axle ratio would have to be reduced by some 30% for equivalent engine speeds, resulting in an ineffi-cient hypoid-gear axle design. Would these reasons portend transmission designs with even two overdrive gear ratios?

The word "overdrive," perhaps fittingly, has captivated the motoring public. It sells transmissions, if not automobiles (i.e., the current ads of "GM—The Car Company in Overdrive").

In overdrive, the output is geared to turn faster than the engine for fuel economy pur-poses, as already noted. Furthermore, the lower engine speeds are conducive to lower

engine and passenger compartment noise levels. This in turn demands quiet gearing while operating in overdrive, well designed and accurately manufactured.

Years ago, I evaluated prototype transmissions for acceptable gear noise and/or whine, both in the laboratory and in the vehicle. (This was and continues to be a challenging assignment. The objective was to verify that the new gear design had quiet gear-tooth geometry when produced with high-production manufacturing equipment.) Since then, the passenger compartment of automobiles, as well as that of many trucks, has become even quieter, a concession to modern driver expectations for a quiet vehicle. Most find the slightest squeak and rattle, transmission gear noise, or gear shift noise objectionable. James and Douglas [5-1] advise that "Noise reduction is increasingly becoming an important subject for transmission engineers." However, as noted next, perhaps noise is only one of many considerations that must be addressed before releasing a new gear design.

Before we can address our goal for an optimum geared transmission, particularly for an automotive vehicle, we should examine twelve considerations of gearing arrangement applications and their effects on gear design:

1. The vehicle application
2. Packaging space
3. The gearing arrangement
4. Torque multiplication range
5. "Elegant," appropriate ratio steps
6. Shift controls and logic
7. Interface with engine controls
8. Lubrication
9. External transmission controls
10. Minimum gear noise and/or whine
11. Gear durability
12. Manufacturing facility requirements

These considerations will next be discussed in more detail.

The Vehicle Application

This criterion for a transmission includes performance and durability factors. In the case of performance, manufacturing economics often dictate extending a gearing design over more than one engine or engine modification or vehicle niche, with the resulting compromise. In the case of durability, because redundancy is not a viable option, the design analysis and testing must reflect the projected vehicle life operating cycle.

Other assumptions applicable to a transmission designed for a particular vehicle application include the following:

• Reasonable production volume (and necessary manufacturing facilities)

• Minimum internal and external controls components

• Minimum or no required maintenance

• Adequate transmission cooling

• Provision for servicing and technician training

• Modular construction

When a transmission is extended across vehicle niches (e.g., an economy vehicle version, base vehicle, and performance vehicle), certain compromises come into play, as shown in Table 2.

TABLE 2
COMPROMISES OF EXTENDING THE BASE
TRANSMISSION TO OTHER VEHICLE NICHES

Attribute	Economy Vehicle	Base Vehicle	Performance Vehicle
Startup Performance	Greater weight than optimum	Design	More usage in the lower gears than design
Packaging Efficacy	Larger than optimum	Design	May need additional transmission cooling
Durability	Greater than necessary	Design	May be limited
Fuel Economy	Compromised by unnecessary weight?	Design	Not a consideration?
Shift Smoothness	Complexity	Design	Compromised

The attributes shown in Table 2 suggest the long-time trend toward cultivating transmission brand name recognition such as "Powerglide" or "Powertorq," as offered by Hafemeister.[5-2]

Packaging Space

Transmission packaging configuration is dictated by the vehicle concept, whether front- or rear-wheel drive, as well as the internal gearing arrangement. Many other factors beyond gearing also come into play, such as the suspension of the powertrain in the vehicle chassis, driveline connections, and arrangement of ancillary components such as coolers and filters.

The Gearing Arrangement

The gearing arrangement represents the heart of the transmission. It establishes the powertrain experience for the driver, if not the vehicle experience. Beyond the layshaft gear arrangement common to the manual transmission, the simple planetary gearset has been the basic gearing building block for automatic transmissions. To build on the simple gearset, there are many possible gear combinations of planetaries. Two of the best known are the Simpson and the Ravigneau. Although a basic planetary gearset had been employed in the Ford Model T, eventually discontinued in 1927, it took a creative engineering consultant, Howard W. Simpson [5-3], to conceive and patent some 42 different viable planetary arrangements. He firmly established the three-speed arrangement using a common sun gear and two identical ring gears. Named after him, this arrangement was patented in 1955. Beginning with the Chrysler Torqueflight transmission, the Simpson set was later adopted by Lincoln, Mercury, and Thunderbird, and subsequently by Cadillac and other General Motors vehicles— The Big Three."

In general, a planetary gear transmission is composed of combinations of three basic elements: sun gear, ring gear, and pinion gears mounted in a carrier. The elements are combined and driven via suitable friction elements to produce ranges of underdrive, direct drive, overdrive, and reverse gear as programmed.

The example shown in Figure 4 illustrates the basic clutch and brake connections to any gearing arrangement, as well as the inertia forces acting on it. This example also illustrates the basic in-line arrangement between input and output, but any other arrangement may be similarly treated.

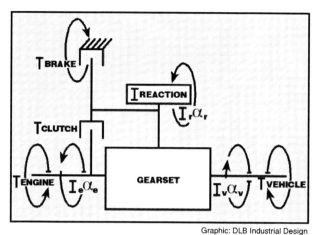

Graphic: DLB Industrial Design

Figure 4. Basic gearset inputs and reactions. (Source: Reference 5-7, Figure 20, page 89.)

What criteria affect the choice of gearing arrangement, and how are they determined?

• Vehicle performance objectives

• Engine torque-speed characteristics

• Transmission ratio spread, number of gear shifts

• Mechanics of ratio change: shifting, infinitely variable, and variation of the two

• Manufacturing facilities

When vehicle performance objectives have been established and an engine and its operating power curve selected, work may begin to develop the new automatic transmission design. Assuming that a planetary gear type has been selected because of available manufacturing facilities, the gear arrangement may be determined. (Other types of ratio-changing transmissions will be discussed later.)

Planetary gear arrangements may be analyzed using a kinematics approach, a lever analogy, or a graphical method as described by Park and Oh [5-4] and others. Utilizing the linearity of the speed relationships among the gear elements, they developed a convention for determining feasible multiple-ratio arrangements. For example, there are 1,249 possible combinations using two simple planetary gearsets, and not all of them are feasible. Park and Oh applied their convention beginning with a basic simple planetary gearset, then expanding the concept to four-speed and five-speed

gear arrangements. The convention was then expanded to create a new six-speed front-wheel-drive arrangement, attractive for its capability to be controlled by only five friction elements, as shown in Figure 5.

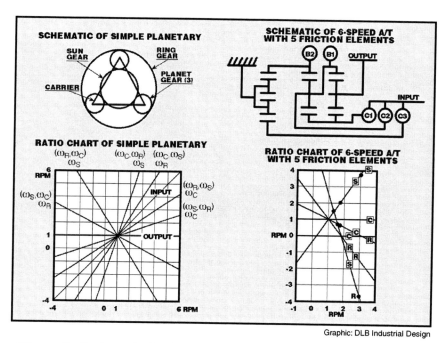

Graphic: DLB Industrial Design

Figure 5. Ratio analysis of a simple planetary extended to a six-speed arrangement. (Source: Reference 5-4, Figures 1, 2, 15, and 16.)

Common practice is to design the planetary carrier with three or four planet gears, although some units have up to six planets for torque capacity. In any event, gear-tooth contact is subject to several sources of error, including planet pin location, tooth thickness variation, and radial misfit due to bearing misalignment as well as component distortion. Seager [5-5] found three planets to be optimal from the standpoint of dynamic contact interaction. He determined that when a planet carries more than its share of load due to elastic deflection rates and dimensional error, the gear-tooth load could be as much as 1.5 times its fair share. Gear whine and durability issues brought on by uneven gear-tooth contact, internal thrust washer wear, and bearing degradation can result.

In addition to establishing gear ratios with the appropriate number of gear teeth, complete gear design provides optimum gear-tooth profiles for durability and noise reduction. To ensure optimum gear-to-gear contact, precise bearing mounting is required for all the gears and shafts to prevent detrimental deflection under load. Also, the planet gears must be mounted on precision needle bearings. To minimize friction losses, the thrust reactions created by the input torques and gear helix should be balanced and contained with adequate thrust bearings in all the operating gears.

Although gear design techniques cannot be adequately covered here, enough has been described to appreciate that less compounding of a gear arrangement is advantageous. That is why simple gearsets, alone or in combinations such as the Simpson, have been popular. On the other hand, compounding, such as the Ravigneau gear arrangement, has been popular in some transmissions since around 1950. That arrangement employed both long and short double planet gears mounted in the carrier assembly, highly developed to minimize deflections to ensure ideal tooth-to-tooth contact and gear durability.

A recent rear-wheel-drive transmission introduction, the ZF six-speed HP26 (Lepelletier-type gear arrangement) has expanded the versatility of the Ravigneau-type gear arrangement to six forward speeds by adding a simple planetary with a creative arrangement of friction control elements. Lepelletier [5-6] has developed an eloquent logic applicable to any multiple planetary gear design by utilizing the basic three- and four-member gearsets to exclude redundancy of components and potential issues such as damaging speeds or double shifting. Adherence to this logic profitably caps the number of planetary gear speeds to six for most vehicles, for the next greater optimum number of speeds is eleven.

Torque Multiplication Range

The required range of torque multiplication is necessarily closely entwined with the performance characteristics demanded of the vehicle. Years ago, units were sold with overall startup ratios, including the axle ratio, of 6:1. Now, 6:1 startup ratios are common for the transmission alone, with additional multiplication provided by the axle and/or final drive. This growth in ratio range has been driven primarily by the global demand for fuel economy.

Interestingly, the growth in range has been provided primarily by additional transmission gear steps with the usual, if not slightly lower, torque converter stall torque ratios. In recent designs, the stall torque ratio is somewhat reduced in favor of increased torque converter operating efficiency and torque capacity. The lower stall ratio somewhat offsets the efficiency loss if the torque converter fluid circuit is "squashed" in axial length for transmission packaging reasons.

"Elegant," Appropriate Ratio Steps

Elegant geometric or harmonic ratio steps of multiple-step transmissions contribute to a more pleasing driver experience by providing a continuous level of function, particularly during the acceleration stage, although not necessarily the optimum performance. In fact, opposing considerations are in play. On one hand, a sequence of small and equal gear-to-gear ratio steps results in lesser disturbances during shifts but requires more gear changing within a given ratio range. On the other hand, unequal steps to complement the engine torque curve or gear design may provide better performance.

Methodology exists to determine and analyze all possible gear configurations that meet the number of gear steps desired, as shown for one method in Figure 5. However, there are other considerations in the selection of a gearing configuration, such as mechanism complexity, controls complexity, and weight, as well as existing manufacturing technology—not to mention cost.

Today, the transmission designer has less room with which to work, in profile or length, given the ongoing demands for greater passenger space and luggage space within a given vehicle class.

As stated in Chapters Two and Three, the various front-wheel-drive and rear-wheel-drive configurations are dictated by vehicle class, image, and packaging configurations. Significant compromise and complexity can result if the design of the transmission must be changed from the optimum to meet vehicle mandated road clearance or wheel-to-wheel space.

Is there an ideal gear ratio step? A check of several recent production automatic transmission designs yields steps ranging from 0.59 from first to second gear down to 0.17 between the next-to-last gear to the final drive gear.

Shift Controls and Logic

Customer controlled functions, such as position of the transmission manual shift lever, manual power shifts, and hill-brake demands, establish the driving mode. To take optimum advantage of a particular gearing arrangement design, data from transient vehicle conditions such as vehicle startup, stop-and-go-driving sequences, driver kick-down demands, and uphill/downhill conditions must be rapidly processed within the controls algorithm. To accomplish this more accurately, the industry is progressing to 32-bit microprocessors from the previously standard 8-bit designs. Responding to these transient conditions, the CPU rapidly interprets data representing engine throttle level, torque converter turbine speed, and output shaft speed, which corresponds to vehicle speed.

Monitored fluid control pressures and clutch engagement pressures are the final link in the chain, with compensation for variations of fluid operating temperatures. Shift controls considerations are advanced in Chapter Six.

Interface with Engine Controls

For optimum shift engagements, engine throttle conditions must be closely integrated with transmission control logic. Accurately de-torqueing the engine through fuel or engine spark control during power transfer from one gear to the next is one key to providing a smooth shift feel.

Lubrication

Although almost an obvious requirement, the proper lubrication of a transmission gear system involves the consideration of mechanical, dynamic, and ambient conditions. This challenging task is affected by several conditions to which the gear system is exposed. For example, after a vehicle is left standing for a long time, lubricating fluid tends to drain down to the sump area from critical surfaces and must be replaced under pressure quickly upon startup. As another example, if a vehicle is to be towed (with the engine not running) for a long distance, lubrication must be provided to the rotating transmission components. As a different example, high-performance driving may incur high stresses and high rotational speeds in components that must be lubricated. Also, sump fluid level is critical and can rise with increased temperature. The rotating components cannot simply be flooded with fluid. That would result in transmission churning losses and eventual degradation of the lubricating and chemical attributes of the fluid.

Current practice leans toward specifying no fluid changes for the life of the vehicle unless it is driven to unusually severe operating conditions, generally associated with trailer towing. Close control of operating fluid temperature is one feature that makes extended fluid life possible.

External Transmission Controls

The span of external types of transmission controls now runs from various mechanical linkages to remote electronic activation for selection of gear range, control of performance options, and even the "park" option. Beyond safety and redundancy considerations of the external control, the type of mechanism is somewhat dependent on the vehicle application except that the advent of electronics permits much more flexibility in the design, as well as for additional driver options.

Minimum Gear Noise and/or Whine

Although mass production gear-manufacturing experience is invaluable to ensure a quiet gearset transmission, the fundamental gear arrangement must be sound for a quiet end product—no pun intended.

Gear Durability

Minimizing stress design loads and relative surface speeds are the conventional techniques to enhance durability. However, the marketing appeal of a greater number of transmission gear ratios to obtain broader operating range begets attention to the finer points of gear design. This has been evidenced in transmission introductions of the past few years. Balancing of internal thrust loads, ample gear-tooth contact ratios, ample bearing supports, and ample fluid lubrication all have priority over the desire to minimize inertia mass, unit weight, and manufacturing complexity.

Manufacturing Facility Requirements

Transmission gear manufacturing is a highly specialized field. Today's gears are far from the "cog wheels" that they might have been jokingly called years ago. Often, the introduction of a new transmission, whether manual or automatic, heralds the number of forward gears that it sports for performance, if not marketing purposes. However, the destiny of gearing is to remain unseen and unheard, particularly in the shadow of the omnipresent engine, which in reality is a severely limited power source without some kind of transmission.

Mass-production, high-precision gear manufacture involves accurate processing such as broaching, gear rolling, fine stamping, sintering, and finishing. These mass production processes have greatly superseded the conventional gear cutting processes such as hobbing and shaping. After gears are finished (dimensionally), they usually are heat treated because they are designed to operate at stress levels to be as space efficient as possible. Heat treatment processes have developed into a fine art, taking full advantage of the intrinsic metallurgical properties of the gear material. Conducted to produce minimum distortion, heat treatment adds the required gear durability without sacrificing quiet gear operation. Selection of a specific heat treat process, relative to the steel specification, is amply covered in the literature. Much has been published on manufacturing considerations affecting gear design. A. Hardy's paper on this subject [5-7] is a useful reference.

All of the previous discussion translates to the need for highly developed precision manufacturing equipment. Thus, to launch a new transmission with a different gearing arrangement, it is always desirable to adapt much of the existing precision equipment to minimize launch and purchasing costs.

To meet the varying demands of the automotive market, the manufacturing industry is going through many changes, such as assimilating flexible manufacturing for transmissions and their components. This flexibility enables rapid change to accommodate various types of gear fabrication, as well as the assembly of a range of clutch capacity components and electronic control combinations for the applicable transmission–engine–vehicle installation.

References

5-1. James, B., and Douglas, M., "Development of a Gear Whine Model for the Complete Transmission System," SAE Paper No. 2002-01-0700, Society of Automotive Engineers, Warrendale, PA, 2002.

5-2. Hafemeister, G., "Trend Towards Cultivating a Transmission Brand...," *Ford ATEO Transmission Review*, Vol. 10, No. 1, March 2002.

5-3. Interview with Bruce Simpson, son of the late H.W. Simpson, October 24, 2002.

5-4. Park, J., and Oh, J., "Analysis of the Gear Shift Mechanisms by R-R Chart Method," SAE Paper No. 2001-01-1163, Society of Automotive Engineers, Warrendale, PA, 2001.

5-5.   Seager, D.L., "Load Sharing Among Planet Gears," SAE Paper No. 700178, Society of Automotive Engineers, Warrendale, PA, 1970.

5-6.   Lepelletier, P., "Advanced Planetary Step Ratio Transmission Technology," SAE Emerging Transmission Technologies TOPTEC Symposium, August 12, 2003, Society of Automotive Engineers, Warrendale, PA, 2003.

5-7.   Hardy, A., "Manufacturing Considerations Affecting Gear Design," in *Design Practices: Passenger Car Automatic Transmissions*, AE-5, Society of Automotive Engineers, Warrendale, PA, 1962.

Chapter Six

The Clutches of Change

Next to the design of gearing arrangements, many transmission engineers agree that the design and arrangement of plate clutch and brake configurations are most significant. Each clutch in an automatic transmission is subjected to many engagements and disengagements at varying loads and speeds over the life of a vehicle. However, the clutch must essentially retain its original design friction and shift characteristics to ensure that shifts perform consistently as calibrated.

An optimum geared transmission should contain the fewest number of internal shift elements required to perform all the shift operations.

The most common type of these shift elements is the friction clutch member. Its distinct task is to connect adjacent components that are rotating at relative speeds. The friction clutch is disengaged when apply pressure is released by an appropriate signal during a shift sequence. Without getting into the intricacies of clutch design, suffice it to express the basic formula for the torque capacity T of a plate clutch as

$$T = n\mu F R_m \qquad\qquad (4.1)$$

where

n = number of clutch engaging surfaces
μ = coefficient of friction
F = apply force
R_m = mean radius of the friction surface

The reason for identifying this basic equation is to point out the engineering problem that each one of the four terms represents. Multiply this problem for five or six separate clutches, and there could be a literal permutation of issues. Therefore, to reduce design and manufacturing complexity, the clutch engineer will strive to "commonize" clutch plate sizes. This requires enlightened design. In any particular transmission gear arrangement, each clutch must absorb different energy levels at different times

and for a different number of engagements in its lifetime, as shown by Winchell and Route.[6-1]

There are some nuances for each of the four terms n, m, F, and R_m.

Although the number of clutch plate engaging surfaces n in a transmission assembly is fixed, the share of the load carried by each plate surface may vary. This share depends on the location of the plate in the clutch pack, the physical structure of the friction surface, the design of the separator plates, the splined connections, and other factors. To promote load distribution, it is desirable to have as few plates as possible. This, of course, conflicts with the objective of using the smallest possible diameter clutch plates to minimize the overall transmission profile.

Elaborate fabricating techniques can produce and control the desired engaging surface coefficient of friction μ by means of various mixes of organic and metallic materials. The type of grooving in the friction surface also can be a factor. The coefficient of friction can be checked using a standard friction testing machine (SAE J286, Clutch Friction Test Machine Test Procedure). However, in an actual transmission installation, the effect of fluid properties, fluid temperature, and friction material stability with shift frequency may vary the coefficient from the design value. The result is unintended shift feel variation from the design level.

The mean effective friction plate radius R_m is used in calculating clutch capacity. It may be affected by operating changes in the friction surface, plate deflections, and clutch assembly distortions.

The preceding qualifications relate to clutches (or brakes) when applied or disengaged. Upon disengagement, new considerations, such as avoidance of drag and vibrations, may apply.

The properties of clutch friction facings are accurately controlled to ensure consistent shifts. Facing materials, textures, and configurations, such as grooving of the clutch face, are highly developed for shift control, durability, and, when disengaged, even low-drag characteristics. For some applications, the plates are waved in a controlled height for smoother engagements and easier disengagements.

What may cause premature clutch deterioration? There are many possible answers. Beyond refinement of the clutch surface characteristics, the operating environment within the clutch assembly plays an important role. Design of the clutch apply piston, piston release springs, clutch plate lubrication, and the controls logic are all critical. In his paper on multiple-disk clutches, Hilpert [6-2] notes that a clutch can even destroy itself due to gyroscopically induced loading.

Detailed analysis involving the modeling of applied clutch torque, friction correlated with the instantaneous temperature rise, and inertia moments of the individual components can be used to project clutch life by summation of the work accomplished per gear-to-gear shift over the customer life cycle.

Clutch piston return springs and plate cushion springs have a subtle but important role in controlling shift characteristics and clutch life. Multiple round wire coil spring assemblies are the most common because they provide better control of clutch disengagements. However, disk and wave springs are employed for cushioning clutch engagement where space is at a premium. Park et al. [6-3] found that with proper selection of the return and cushion springs, the clutch overall axial length can be reduced by almost 15 mm (0.6 in.). The weight of the clutch assembly also is reduced, which is a decided advantage. However, the spring design decision must be made in light of other design and manufacturing considerations. For example, it is significantly more difficult to control the characteristics of a compact wave spring in the manufacturing phase, compared with multiple coil springs, commensurate with acceptable spring durability.

In summary, the overall performance of a multiple-plate clutch involves many engineering considerations, including centrifugal hydrodynamic piston pressure balancing and optimum solenoid and electronic control unit (ECU) self-compensating features. Minimizing the drag of a disengaged, open multiple-plate clutch is dependent on low relative speed, clutch pack clearance, friction groove pattern, and plate flatness, as well as the ability to evacuate all lubricating fluid.

Issues that apply to the common multiple-plate clutch have been described thus far. However, there have been other popular clutch designs. For example, cone clutches have been employed where relatively large capacity is needed in a limited space. Also, brake bands, which are applied by servos tightening the band over an appropriate drum, have been used extensively in the past. Their excellent brake capacity in a relatively small space made them popular, although their unidirectional friction characteristics have made them more difficult to control.

The so-called one-way clutch (OWC) is a mechanical form of a clutch and will be discussed here. As its name suggests, the OWC prevents the rotation of an element in one direction relative to another element or with respect to the transmission housing. It releases and free-wheels instantly at the cessation of reaction torque during a shift event. As used in a transmission, it is a highly developed and sensitive mechanism. The mechanical OWC has been employed in most transmissions since around 1950. However, to reduce complexity, space, and cost, serious efforts to eliminate the OWC in the shift process continue to be made since the introduction of the 1989 A604 Ultradrive with direct acting, pulse-width-modulated (PWM) control solenoids controlling the application of plate clutches.

The OWC has taken various design forms over the years. Initially, radial-action sprag-type clutches developed by Borg Warner were used, and these underwent much development to reduce drag and improve durability. Also, roller clutches were introduced for their relative ease of manufacture. More recently, ratcheting axial blade one-way clutches, developed by Means Industries, have replaced the roller clutch in some torque converter reactor and gear reaction applications. In gear-shift reaction applications of the axial ratcheting type design, it is important to take all the inertia forces into account.

References

6-1.   Winchell, F.J., and Route, W.D., "Ratio Changing the Passenger Car Automatic Transmission," SAE Paper No. 311A, 1961 SAE International Congress and Exposition of Automotive Engineering, Warrendale, PA, 1961.

6-2.   Hilpert, C.R., "Gyroscopically Induced Failure in Multiple Disk Clutches, Its Causes, Its Characteristics and Its Cures," SAE Paper No. 690066, Society of Automotive Engineers, Warrendale, PA, 1969.

6-3.   Park, D.H., et al., "Optimum Design of Return and Cushion Springs for Automatic Transmission Clutches," SAE Paper No. 2001-01-0870, Society of Automotive Engineers, Warrendale, PA, 2001.

Chapter Seven

Transmission Controls

To take full advantage of the multiplicity of gear ratios that are provided in an automatic transmission, it is necessary to pay attention to all aspects of the controls system. This also applies to the infinitely variable-type transmission with its special features.

The objective of a controls system, perhaps as its name implies, is to precisely convey the demands of the driver, within the constraints of the vehicle and the roadway, to the powertrain system.

The usual goal of the driver of a vehicle is to travel from point A, the vehicle startup location, to point B, the driver's destination, effortlessly and with a sensation of being in control. For the transmission engineer, this can present a substantial discord ranging from offering complete driver control by shifting manually, as with the conventional manual transmission, versus dependence on preset automatic controls to respond to driver and road performance demands. Both extremes can involve compromises—in driving comfort, vehicle performance, and vehicle fuel economy. With that said, it is inevitable that today's driver be afforded the opportunity to either shift manually or to leave the gear ratio/engine speed decision up to the ubiquitous controls. Developments in electronic and computer controls are making that a reality.

What are the basic functions of an automatic transmission control system?

- Schedule transmission shifts to complement engine characteristics

- Provide the optimum gear ratio and speeds on operator demand

- Provide a safe transition to reverse gear operation

- Provide assured manual access to the range of options available and shown in the instrument panel shift quadrant (e.g., "Park," "Reverse," "Neutral," "Overdrive," "Drive," and "Manual Low")

- Provide for torque converter lock-up, if applicable, to eliminate hydrodynamic slip under cruising conditions

- Provide for engine braking assist in low gears, as on long mountain grade descents

The controls system generally includes the following basic components needed to execute the preceding functions:

1. Hydraulic valve body incorporating cored bores for various valves and related interconnecting fluid passages and appropriate orifices

2. Hydraulic shift control valves
   a. Spool types, pop-offs, reed, etc.
   b. Valve control springs
   c. Check balls for fast fluid exhaust

3. Regulator valves for analog control of pressure or flow

4. Electronic solenoids to actuate valves to execute shifts
   a. On–off type
   b. Variable-force type
   c. Modulated type, etc.

5. Hydraulic servos to apply and release bands, if applicable

6. Accumulators to soften upshifts

7. Electronic control modules to process throttle position servo data, and engine and turbine speed data, and to signal appropriate events

8. Sensors for transmission range position, engine throttle position, engine mass flow, engine speed, transmission output speed, and fluid temperature

Utilizing adaptive real-time control of clutch capacity, integrated with ratio-change feedback, transmission clutch-to-clutch upshifts and downshifts are smoother and more consistent.

With the introduction of electronics and development of electronic control modules, the technology has advanced to the point where gear change can be accomplished by applying a clutch while disengaging another, obviating the need for a one-way clutch (OWC) in many applications. Clutch-to-clutch shifting simplifies the valve body by adding electronics to meet the basic objective of synchronizing the oncoming clutch apply circuit with the exhausting clutch circuit. This permits greater flexibility in control strategy by employing computer software rather than fixed hydraulic circuitry. There is less mechanism complexity and weight, as well as the opportunity for cost reduction.

To ensure consistently smooth clutch applications at all temperatures without tie-up or flare-up, the controls must be more definitive when an OWC is not employed. The dynamics of the shift solenoids, clutch system, and accumulator system must be controlled together with the pressure regulator system.

To synchronize the disengagement of one clutch and the application of another, component speed data are generally processed through the control algorithm utilizing robust sensor technology. For improved synchronization, Bai et al. [7-1] have demonstrated the advantage of introducing clutch pressure feedback so that the disengaging clutch pressure is always above the pressure at which slip could occur independently of the fill time of the engaging clutch. This circuit is illustrated in the schematic of Figure 6. Bai rightly points out that as the oncoming clutch reaches its critical engagement, the exhausting clutch capacity must become zero. To prevent flare-up, the exhausting clutch is set above its critical capacity before the oncoming clutch reaches its critical engagement capacity.

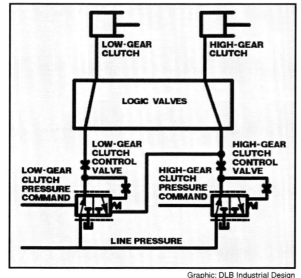

Graphic: DLB Industrial Design

Figure 6. Clutch-to-clutch shift control using oncoming clutch shift pressure in the synchronization algorithm. (Source: Reference 7-1, Figure 1(b).)

Without getting into the design specifics of individual hydraulic/electronic controls, some basic requirements include keeping the mass of the hydraulic valving as low as possible for more rapid response, providing sufficient flow circuit area, and minimizing potential areas of leakage. It is important that the electronic circuit provides for external shift logics testing, as well as servicing the solenoid valves if necessary.

The discussion of controls would not be complete without commenting on the important task of communicating the demands of the driver to the transmission and engine. The old methods of mechanical linkage and cable connections for gear shift position, park engagement, engine throttle position, performance, and economy control options are gradually being replaced with more precise and flexible electronic systems loosely known as shift-by-wire. The need to continuously integrate the data from many sensors to meet the demands of today's vehicle has, in effect, created a separate new field of controls.

References

7-1.   Bai, Shushan, et al., "Development of a New Clutch-to Clutch Shift Control Technology," SAE Paper No. 2002-01-1252, Society of Automotive Engineers, Warrendale, PA, 2002.

Chapter Eight

Performance Attributes

The performance characteristics of a vehicle with an automatic transmission will be addressed initially, although the delineation between it and a manual transmission has continued to narrow with new technology. Assuming a typical three-element torque converter, an early objective for a new design is to provide for an optimum "match" between the engine and the transmission.

The conventional eight-cylinder engine is characterized by its torque peak at relatively higher speeds, compared with the somewhat flat torque curve of most four-cylinder and six-cylinder engines. The torque converter especially complements the eight-cylinder engine by augmenting the vehicle overall startup torque.

Optimum vehicle performance is approached with the blend of several criteria that impact the driver, such as initial startup acceleration, demand-acceleration from various cruising speeds, and fuel economy. El-Sayed and Song [8-1] describe a mathematical optimization modeling procedure for vehicle acceleration. Acceleration and velocity are projected from net tractive effort at the vehicle wheels, including the effects of inertia and weight transfer from the front to the rear wheels. Similarly, optimum steady-state fuel economy was modeled based on engine brake specific fuel consumption (BSFC) data and the net road load power required at the wheels.

El-Sayed and Song rightly concluded that transmission gear ratios, axle ratio, position of the vehicle center of gravity, and especially the weight of the vehicle are important parameters. Understandably, the optimum vehicle was found to require design compromise of transmission ratios to meet both acceleration and fuel economy criteria.

Although the aforementioned analysis was based on a vehicle with a five-speed manual transmission, a similar reiteration procedure with a torque converter transmission or continuously variable transmission (CVT) will provide similar useful projection of vehicle performance.

In the development stage, transmission efficiency and other performance characteristics may be accurately measured by means of double electric dynamometers. Also, techniques for measuring the performance of a transmission installed behind an engine in a dynamometer setting are available, although these techniques have limitations. However, with the development of microprocessors and modern instrumentation, it is now possible to obtain a good indication of transmission performance characteristics, as well as engine torque characteristics under vehicle transient operating conditions.

For example, the converter/transmission K curve, discussed here previously, may be used. Its values are relatively independent of engine torque at "stall" or at any other specific speed ratio, by nature of the fixed design of the torque converter blading. Therefore, an indication of full throttle as-installed engine torque may be readily obtained by measuring engine revolutions per minute while briefly depressing the accelerator to the floor with the vehicle brake engaged. The resulting stall speed gives a measure of engine torque applying the relationship $T = [N/K]^2$, or engine torque may be read directly from a typical test curve, such as the curve shown in Figure 7.

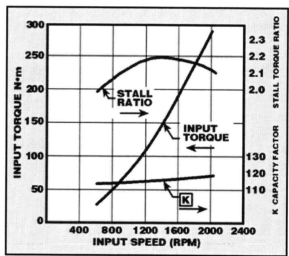

Figure 7. Engine torque versus speed with transmission output stalled. (Source: SAE J643, Figure 4.)

Graphic: DLB Industrial Design

Returning to the discussion of transmission performance, conducting the following test procedures on a sample unit will provide an excellent basis for comparison to other types of transmissions. Projections of vehicle full-throttle performance thus may be developed by inputting relevant engine/driveline/vehicle physical characteristics to computer projection software. The test procedures are as follows:

SAE J643   Hydrodynamic Drive Test Code

SAE J651   Passenger Car and Truck Automatic Transmission Test Code

Figure 8 illustrates automatic transmission parasitic losses in direct gear with the output shaft disconnected. This type of dynamometer test can disclose useful information about the unit, such as total clutch drag losses, gear churning losses (as applicable), and friction losses, as well as pump torque. Of course, a test for these parasitic losses in low gear or in any other gear would yield different results, depending on which gears are in operation and which clutches are engaged.

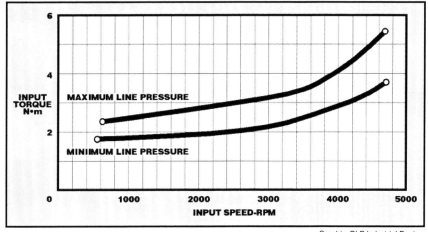

Graphic: DLB Industrial Design

Figure 8. Transmission parasitic losses. (Source: SAE J651c, Figure 7.)

In addition to the conventional planetary transmission, other types such as belt drives, toroidal drives, hydrostatic, the conventional manual, or the dual clutch transmission may be similarly tested and the parasitic losses compared. The controls must be preset in the test to duplicate vehicle operating conditions.

In the preceding example, the transmission parasitic loss at 2400 rpm ranges from 2 to 3 N•m (2.7 to 4 ft-lb), depending on line pressure. The road load requirement of a vehicle is approximately 20 N•m at that driveshaft speed. Therefore, transmission parasitic losses would represent about a 10% additional engine requirement. Reducing these parasitic losses requires enlightened design of all the contributing internal transmission components, such as the clutches, the gearing, and the pump.

The remaining standard tests (not illustrated here) in the aforementioned procedures involve performance runs in various operating modes, such as "Drive" (conducted at 150 N•m input torque for comparison purposes), road load, and coast drive conditions. The runs are all conducted under steady-state conditions so that if the dynamometer data were to be used in projecting vehicle performance, inertia effects would have to be taken into account. To conduct the steady-state tests in various gears and loads, the transmission controls are specially locked in position to prevent automatic shifts. Because the test is conducted at high loads and steady-state conditions, provision for external cooling must be provided to maintain transmission fluid temperatures at actual customer operating levels or maintained at the SAE test standard 100°C (212°F) temperature.

References

8-1.   El-Sayed, M., and Song, D., "Automotive Performance Optimization," SAE Paper No. 980825, Society of Automotive Engineers, Warrendale, PA, 1998.

Chapter Nine

Transmission Power Needs

As is evident from discussions in the preceding chapter, a transmission represents the integration of torque multiplying elements, a system of clutches, a controls system, a power support system, and a structure to encase it all. The power support system comprises a hydraulic pump to provide regulated fluid pressure to the controls system and DC electric power to the computer units and actuating solenoids.

The discussion in this chapter pertains primarily to the hydraulic pump system, which can play a significant but sometimes overlooked role in the performance of a transmission. The pump system provides fluid under high pressure, taking power from the engine. The pump outlet pressure must be regulated to circulate fluid through a torque converter and heat exchanger, fill the clutches, actuate the valves, and lubricate the many bearings and moving components. The high pressures and flows required to actuate continuously variable transmission (CVT) drives can be significant. Because of its power consumption, the design of the hydraulic fluid system demands careful attention.

To minimize power consumed by the pump system, early automatic transmissions were fitted with two pumps: (1) a primary pump driven at engine speed, and (2) a smaller secondary pump driven at output speed which took over during cruise conditions. Noteworthy is that this secondary pump provided the capability for vehicle push start. Reflecting the increased reliability of modern the engines and changes in vehicle body design, this secondary pump was eliminated. Since then, fluid pressure is provided by a single engine-driven pump incorporating provision for fluid recirculation and other features to reduce losses through system electronic controls. The fixed displacement pump must be large enough to provide adequate flow capacity at low engine speeds, but it is usually too great under high-speed cruise conditions. To offset the resulting power loss, an internal variable displacement pump may be substituted, but the added mechanism threatens cost-effectiveness of the pump system. Providing an external pump driven by a variable-speed electric motor may be an alternative.

The design of the engine-driven positive displacement pump and integrated control system has undergone extensive development within the competitive constraints to provide high volumetric efficiency at low noise level. The pump system must operate in both cold and hot vehicle environments. It must operate at a wide range of pressures and flow requirements to apply the clutches and lubricate the precision components. Because transmission fluid is not incompressible, achieving pump efficiency and low noise levels can benefit from detailed time-dependent computer analysis followed by development effort.

Significant pump design considerations are as follows:

- Component robustness.

- Favorable pump inlet and outlet porting geometry to minimize cavitation losses (particularly at the inlet port).

- Precise control of running clearance between the gear and its housing for maximum pump output and efficiency with minimum fluid drain-back. This is best accomplished using the same material, preferably ferrous, for both the pump body and the pump gears to rule out expansion differential due to temperature.

Haworth et al. [9-1] reported that "eliminating transmission pump system losses could yield between 1% and 2% composite-cycle fuel economy gain." Beyond paying attention to good inlet port geometry, their analysis suggests that the porting area should be sufficient to limit inlet-side fluid velocities to 15 m/s.

Roethler et al. [9-2] report that the variable-vane type of pump has the lowest efficiency of the five different designs tested because of the relatively greater leakage paths, as shown in Figure 9. However, the lower efficiency can be somewhat offset by reducing the pump displacement at high operating speeds to match pump flow delivery to the transmission flow requirements. Power to drive a fluid pump is essentially proportional to the displacement, the operating pressure differential between the pump inlet and outlet, the pump revolutions per minute, and the fluid velocity. In addition to porting design and optimum running clearances, pump efficiency may be enhanced by recirculation of the return flow back to the pump.

In addition to the variable-vane type of pump design, other common types are the involute gear, hypocycloidal with a dividing crescent, gerotor, and dual stage. The

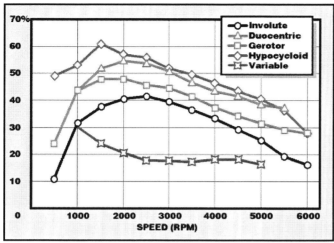

Figure 9. Comparison of overall efficiency of five pump types. (Source: Reference 9-2, Figure 13.)

Graphic: DLB Industrial Design

gerotor pump is preferred for manufacturing reasons because it has no crescent dividing the internal and external gears. Although the design has the potential for less cavity disturbance with its distinctive external and internal lobe running contacts, the controls circuit requires attention to the avoidance of possible resonance resulting from the lower running frequency. A dual-stage pump design, fitted with many front-wheel-drive transmission configurations, scavenges the fluid from the bottom pan to the side pan that houses the pump, regulator, and control body.

Selection of the best pump is determined by the flow and pressure requirements and the transmission configuration; however, attention to the fine details of pump design can make a significant contribution to efficiency, particularly under cruising conditions.

References

9-1.   Haworth, D.C., et al., "Dynamic Fluid Flow Analysis of Oil Pumps," SAE Paper No. 960422, Society of Automotive Engineers, Warrendale, PA, 1996.

9-2.   Roethler, R.C., et al., "A Performance Comparison of Various Automatic Transmission Pumping Systems," SAE Paper No. 960424, Society of Automotive Engineers, Warrendale, PA, 1996.

Chapter Ten

Transmission Efficiency and Internal Component Power Losses

The subject of transmission overall efficiency and internal component power losses has been briefly discussed in the preceding chapters. Projection of efficiency and performance of automatic transmissions with a reasonable degree of accuracy is at hand with available computer programs. The software is particularly useful in evaluating the impact of component modifications, as well as for projecting vehicle performance. Park's analysis [10-1] is a sampling from the literature.

The pie chart in Figure 10 of the relative power distribution was projected for a four-speed front-wheel-drive torque converter transmission running at 2500 rpm in fourth gear and 88 Nm (65 lb-ft) input torque. This example illustrates the distribution of power losses that can exist for a substantial portion of driving time under more or less cruising conditions.

Power Distribution	Percent
Friction elements	32
Oil pump	29
T/F and forward gears	13
(Fluid) churning	10
Bearings	6
Oil seals	5
Planetary (Ravigneau)	3
Bushings	2
Total:	100

In this example, the friction elements and oil pump represent 61% of the total power loss. Of course, this distribution of component power losses will change according to the gear and the different loads and speeds. For example, in low-gear startup, the torque converter is active and contributes some power loss even as it multiplies engine torque.

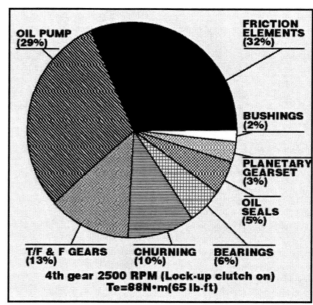

Figure 10. Power distribution. (Source: Reference 10-1, Figure 10.)

Graphic: DLB Industrial Design

Without discussing the reasons for the power loss that each of the powertrain components contributes, a fertile area for economy improvement is disengaging the torque converter from the engine at idle. Power loss exists in the torque converter during vehicle engine idle in drive gear. As previously noted, the engine torque required to drive the impeller when the vehicle brake is engaged and the turbine is held is proportional to the square of the engine speed. Although the resulting power loss is manageable under normal engine idle conditions, the power loss is greater at the higher engine idle speeds occurring during cold start conditions or with the vehicle air conditioning system engaged.

Hayabuchi et al. [10-2] demonstrated that by disengaging the torque converter from the engine, fuel consumption can be reduced by 20% while the vehicle is at rest (e.g., waiting for a traffic light to change). They note that for the typical traffic congestion pattern in Japan, this equates to an overall 5% fuel improvement. Of course, the torque converter was originally integrated in the transmission design to serve the important functions of smooth vehicle startup and engine torsional vibration damping. Therefore, these attributes should be retained with the addition of the idle disconnect feature. To avoid engine flare-up, maintaining partial clutch engagement may be necessary. Also, precise control of apply and release pressures is necessary to ensure a natural feel during vehicle startups and rolls-to-a-stop. Provision to hold the vehicle from rolling backward on grades must be made to replace the hydrodynamic drive capability inherent to the torque converter. To ensure satisfactory clutch life with the idle disconnect feature requires durable clutch friction facings, robust apply piston and backing plates, and control of open clutch plate running clearances.

Under the ongoing mandate to reduce overall vehicle fuel consumption, the neutral clutch feature described previously or some other type of disengaging device probably will eventually become standard equipment.

Rekindling of domestic interest in diesel engine installations for fuel economy gain requires appropriate attention to the design of the automatic transmission. The good low-end diesel engine torque characteristics and excellent fuel efficiency are very attractive. DeHart [10-3] notes that only about 1% of domestic vehicles, of more than 16 million produced in 2002, versus 39% in Europe, are equipped with diesel engines. However, he estimates that each one-mile-per-gallon improvement in fuel economy results in a $130 to $150 penalty for a diesel installation over gasoline, as compared with only a $50 to $100 penalty by upgrading to a six-speed automatic transmission.

References

10-1.   Park, Dong Hoon, "Theoretical Investigation on Automatic Transmission Efficiency," SAE Paper No. 960426, Society of Automotive Engineers, Warrendale, PA, 1996.

10-2.   Hayabuchi, Masahiro, et al., "Automatic Neutral Control—A New Fuel Saving Technology for Automatic Transmission," SAE Paper No. 960428, Society of Automotive Engineers, Warrendale, PA, 1996.

10-3.   DeHart, Kevin, "Diesels and Their Future in the U.S.," May 28, 2003, SAE Executive Management Briefing, Michigan State University Management Education Center, Society of Automotive Engineers, Warrendale, PA, 2003.

Chapter Eleven

Harnessing Noise, Vibration, and Harshness (NVH)

As vehicles become more luxurious, noise, vibration, and harshness (NVH) issues related to the transmission continue to demand more attention in vehicle installations. Whether the powertrain arrangement comprises the conventional front-engine rear-wheel drive, front-engine front-wheel drive, or some other combination, it is most advantageous to address transmission NVH issues in the design stage.

In a rear-wheel-drive powertrain arrangement, the interaction within the entire engine–transmission–driveshaft–rear axle, as well as the vehicle suspension and driveline mounting locations, all play important roles in defining the impact of the transmission on NVH. Engine and transmission mounts must be precisely located to preclude high-amplitude disturbances to minimize noise and durability fatigue issues.

Although the inherent structural rigidity of a front-wheel-drive transmission arrangement tends to isolate NVH issues, it can exacerbate other aspects. The location of the transaxle, close to the driver's feet, is a more direct conduit for telegraphing noise and vibration to the passenger compartment.

What are the sources of these transmission NVH issues? The prominent sources are the gears, the torque converter (particularly when the lock-up clutch is applied), the fluid pump, and the gear-to-gear shift dynamics—perhaps in that order. In layshaft transmissions, gear rattle can be a concern without appropriate damping, to the point of providing dual mass flywheels.

Although the subject of gear noise reduction is too vast to permit a detailed discussion here, certain design approaches predominate. For example, robust design of the gears and gear support structure is essential to minimize static and dynamic deflections and misalignments. These deflections can compromise precise tooth-to-tooth contact and induce gear whine. Often, robust design may run contrary to the ongoing demands for greater gear ratio range, lower profile, and lighter weight. This tests the skill of the

transmission designer. Approaches such as discretely adding a weight, developed through finite element methods (FEM) analysis, to an offending gear as described by Miyauchi et al. [11-1] can effectively reduce offending distortion and resonance.

Gear designers have introduced design features such as gear-tooth counter-phasing, generous contact ratios, and helical gear-tooth modifications to prevent the possibility of gear noise.[11-2] In addition, new metallurgical processes and advances in fabricating techniques such as gear-tooth rolling, fine stamping, and sintering with hard finishing have become accepted in mas production and hold promise for more gains in cost and gear noise reduction.

References

11-1.	Miyauchi, Yuma, et al., "Introduction of Gear Noise Reduction Ring by Mechanism Analysis Including FEM Dynamic Tuning," SAE Paper No. 2001-01-0865, Society of Automotive Engineers, Warrendale, PA, 2001.

11-2.	Route, W., "Gear Design for Noise Reduction," in *Design Practices: Passenger Car Automatic Transmissions*, Vol. 5, 2nd edition, Society of Automotive Engineers, Warrendale, PA, pp. 58–70, 1962.

Chapter Twelve

The Transmission Is "Cool"

Some early automatic transmissions such as the 1950 Ford-O-Matic were air-cooled by means of fins located on the outside of the impeller member, which circulated ambient air through the housing. However, with the rapid growth in engine power to meet the demand for increased vehicle performance, transmission fluid cooling via engine–water heat exchangers became the norm.

In the conventional automatic transmission, heat is generated by several components, such as the torque converter, the fluid pump system, the gearing system, and the clutch/band system, as well as by oil churning that may take place within the confines of the transmission case at high speeds. All of these sources of heat may contribute to transmission power losses. If the temperature is not adequately controlled, degradation of the operating fluid and some internal components can take place. For example, the clutch plate friction facing material can wear to the point of shift malfunction or totally destroy itself if not adequately cooled.

Of significance in this discussion are the design considerations in providing an effective cooling system. The transmission is expected to operate flawlessly under a broad range of vehicle operating conditions and ambient temperatures. At vehicle cold starts and low ambient air temperatures, the fluid is more viscous, creating flow restriction within the heat exchanger circuit. Thermistor controls effectively bypass this restriction to ensure sufficient pressure for clutch application, torque converter operation, and component lubrication.

How many seconds does it take for the fluid to warm up for the transmission to operate efficiently? Of course, it depends on the transient heat developed within the torque converter and perhaps from the heat reverse input to the heat exchanger from the engine radiator. With transmissions and engines becoming more efficient, reliable electronic control of the cooler flow circuits is needed to ensure early warm-up. At the other extreme, fluid can reach unacceptably high temperatures at high ambients and heavy-duty driving at maximum engine power, such as running up a grade. Generally, fluid operating temperatures are limited to 260–280°F (approximately 125–140°C) maximum

by the cooling system. At these higher temperatures, greater internal leakages and related pressure drops within the control system can occur. Temperature solenoids must be strategically located within the system to control fluid temperature for the entire operating range.

As a frame of reference for cooling requirements, Figure 11 is a projection of the relative power loss of a five-speed torque converter automatic transmission at full throttle and at road load vehicle conditions, plotted versus vehicle speed. These losses stem from the torque converter, the fluid pump, the planetary gear system, and other components. The losses represent heat that must be dissipated by the cooling system in energy equilibrium within the transmission environment.

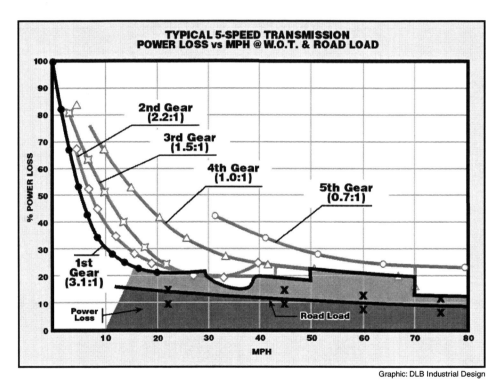

Graphic: DLB Industrial Design

Figure 11. Five-speed transmission power losses.

The cooling system, in addition to its role of cooling the torque converter, provides cooled fluid lubrication to all the internal components, utilizing an external heat exchanger located within the engine radiator. Also, for special vehicle duty such as trailer towing, an auxiliary heat exchanger is available on special order.

The design of the transmission internal cooling valve circuitry is critical to ensure that regulated converter pressure is always available to charge the torque converter. Designs employing a lock-up clutch integral with the torque converter generally include a separate regulator control valve. Internal leakage due to regulator valve wear and valve sticking due to contamination can result in transmission overheating in the field, followed by shift complaints and subsequent clutch failures. This failure mode is generally addressed in the controls design phase by providing an electronically controlled "limp home" mode of operation.

Although the task of cooling the transmission receives considerable engineering attention, it is appropriate to mention here that very low temperatures can be quite detrimental to all transmissions, from an operational standpoint as well as from the impact on efficiency. The torque converter transmission has various electronic and hydraulic circuitry that must function to actuate clutches and maintain lubrication at below-zero startups. Similarly, the traction and toroidal friction continuously variable transmissions (CVT) provide appropriately high Hertzian stresses at the point of contact to ensure that no slip occurs between the pulley and belt or rollers. Specified lubricants must have appropriate additives to ensure controlled friction at very low temperatures, perhaps a limiting operating condition for this type of CVT.

Chapter Thirteen

Transmission Fluid and Lubrication

Chapter Ten identified the various sources of power losses and the need to provide regulated fluid to operate and lubricate the various transmission components.

The fluid employed in the transmission is clearly one of the most important components for satisfactory operation. Fluid development has been ongoing to achieve the lubricity, viscosity range, oxidation resistance, and friction properties required by transmissions. Many physical and chemical properties applicable to a fluid for a particular transmission must be controlled in the formulation process, but that is a topic for a separate discussion. Several automatic transmission fluids (ATFs) are available in the field. Because of their unique characteristics, it is essential that the manufacturer's recommended fluid be used in servicing a transmission.

How is the fluid for a new transmission selected by the transmission engineer? For marketing reasons, it is desirable to use a fluid already available on the market that meets the requirements and has already been proven. However, some transmissions, such as the infinitely variable friction drive arrangement, have their own particular requirements and must be provided for accordingly.

Once the fluid properties are specified, it is important that they not deteriorate during operation over the life of the vehicle. One way to help retain the fluid properties is to maintain the fluid temperature design levels.

Another way to retain the fluid properties—and one that keeps the transmission out of the service garage—is to ensure cleanliness of the entire hydraulic system. Although extensive provisions are generally in place to ensure cleanliness of all the transmission components before and during assembly, some minor contamination is inevitable. This initial contamination, plus particles accumulated with transmission clutch wear, bearing wear, and gear wear during usage must be kept from the critical control elements. Contamination is a significant factor in shift complaints and control malfunctions.

How can the transmission engineer ensure that contamination does not become an issue during the life of the vehicle?

For example, an entire science evolved around the description of the size and constituency of particles contaminating the transmission either due to the manufacturing process or to wear. SAE J1165 discusses a classification procedure based on International Organization for Standardization (ISO) standards.

Droste et al. [13-1] reported on the density of particle distribution from the size of 5 to 50 microns taken from eight different transmissions and different manufacturers. After vehicle test (under about 3,000 miles) and measurement of particle contamination, the results were in the following ranges:

Count	Size of Particles
1–20	50-micron
800–8000	15-micron
More than 50,000	5-micron

It is apparent that contaminants must be contained with appropriate filters, not only in the transmission fluid sump but in the intakes to critical solenoid valves. But what size filter and what size screen should be used? The preceding measurements give some direction toward answering that question. Screening that is too fine results in unacceptable pressure drops; screening that is too coarse negates its intended purpose.

Pump suction-side filters that are too fine can lead to cavitation, particularly with the normal contamination that develops with mileage accumulation. Adding to the problem of filter specification is the controls design trend toward low-force, low-clearance proportional solenoids that demand clean fluids for stick-free operation over the life of the vehicle. As a partial solution, the trend toward additional split-flow side filtration should continue.

References

13-1.   Droste, Timothy A., et al., "Automatic Transmission Hydraulic System Cleanliness—The Effects of Operating Conditions, Measurement Techniques, and High-Efficiency Filters," SAE Paper No. 2001-01-0867, Society of Automotive Engineers, Warrendale, PA, 2001.

Chapter Fourteen

Transmission Sealing Considerations

The leakage of transmission fluid on the garage floor is obviously unacceptable, but it used to be one of the most frequent sources of complaints. To assist in the identification of the source of fluid leaks and to distinguish the fluid from engine motor oil, red dye was added to transmission fluid many years ago.

External fixed seals, such as oil pan gaskets, have undergone extensive development over the years to compensate for machining irregularities, temperature gradient effects when connecting components of different material, fastener characteristics, vibration, and component distortion. That development continues in view of new transmission fabrication techniques and new transmission fluids continually being introduced. Current successful pan-gasket installations utilize hard joint inserts at each bolt location and rubber-type beading along the entire joint.

Internal leakage of transmission fluid can be as serious as external leakage. At various hydraulic operating clearances, fluid leakage is subject to increase with higher operating temperatures. Locations such as control valve to valve body bores, clutch piston seals, and radial clearances between rotating steel shafts and mating bushings are particularly vulnerable. Leakage can manifest itself not only in a greater power requirement, but in a shift controls malfunction because of the inevitable change in hydraulic pressure balance.

To avoid control valve leakage changes with temperature, the valves and mating body bores should be built of similar materials, whether steel, aluminum, plastic, and so forth. Steel or iron combinations were popular for their robustness in the past, but to save weight, aluminum valves and control bodies now have often become the material of choice. With aluminum components, wear and erosion may become an issue over extended mileage, leading to contamination and shift malfunctions as mentioned. Hard surface coatings are commonly applied to key wear areas on control valves to mitigate this issue.

The leakage due to various internal operating clearances can be an issue, as noted. A more daunting issue is the sealing of rotating shafts and components sealed by gaskets exposed to the outside because of the potential for loss of essential fluid.

Almost all transmissions are faced with the task of dynamic sealing with respect to a rotating shaft. Elastomeric materials are the seal materials of choice, but the physical and chemical properties and design of the sealing configurations is a highly developed field in itself. Dinzburg [14-1] outlines the following properties to consider in regard to maximizing seal life and effectiveness:

- Resistance to lubricated wear
- Elasticity
- Magnitude of seal swelling in fluid
- Carbonization
- Initial stiffness
- Dynamic seal pumping ability
- Cold resistance

In a review of these properties, one that stands out is resistance to lubricated wear. Potential wear of the sealing lip contacts with usage must be controlled. Based on bench tests, wear may be empirically defined as

$$W = at^b$$

where W denotes wear, a and b are material compound coefficients, and t is a parameter of time or usage.

To ensure a good seal at the rotating contact, the seal designer utilizes a seal material with appropriate elastomeric properties or provides for this geometric rate of wear by means of some retention device such as a "garter spring." The design of lip seals for rotating shafts is highly developed. One requirement is optimum lip radial loading without excessive drag loss. The structure of the seal lip includes helical grooving on the lip side exterior to the transmission, acting as miniature fluid-filled pumps that thereby add to the sealing efficiency. This feature is particularly adaptable to transmission rotating shaft sealing because one direction—forward rotation—is predominant.

Clutch pistons sliding axially under pressure within their rotating cylinders must be designed for minimum drag force and sealing efficiency. A variety of co-polymerized shapes—including O-rings, so called D-rings, and lip seals bonded to the piston—are used, depending on the application. Bonded lip seal technology has progressed to the point that makes seals bonded to steel pistons preferable for their consistent operation and low drag.

Attention to adequate external and internal sealing design, although not a topic of general interest, is a highly developed and important task in the design of transmissions.

References

14-1. Dinzburg, Boris, "The Selection of Elastomer Compounds Through Correlation of Rubber Properties to Seal Life," SAE Paper No. 2001-01-0686, Society of Automotive Engineers, Warrendale, PA, 2001.

Chapter Fifteen

Transmission Life Cycle and Sustainable Development

In 1996, SAE had taken the initiative to implement policies to promote engineering design for the complete life cycle of all automotive products. Recycling of materials when those materials have exceeded their useful life is encouraged to enhance the environment.

Strides in design continue to be taken to improve the potential for recycling transmission components. The first phase of this goal is to improve the quality and the durability life of the transmission. Reflecting these accomplishments, transmission warranties have gradually improved. From the early warranties of 12,000 miles and/or twelve months, the standard vehicle powertrain warranty has grown to 36,000 to 50,000 miles and/or three to five years, depending on the vehicle. Longer warranties, up to 100,000 miles, are also becoming available in some passenger car lines, but it remains to be seen if these warranties can be economically supported and if it is the start of a trend. Truck installations generally offer warranties of up to 150,000 miles and ten years. Of course, extended warranty coverage for any vehicle can usually be purchased as well.

Maintaining ongoing vigilance over field repairs, responsible manufacturers have developed rapid, extensive dealership reporting systems, beginning with new vehicle introductions and continuing for many years and thousands of vehicle miles for each model. This not only forms the basis for early problem detection and definition but for the control and follow-up of field repairs. One measure that is commonly used is the number of repairs per thousand (R/1000) vehicles at a particular time or mileage interval.

Over the years, the service aftermarket has developed in proficiency by transmission rebuilder organizations such as the Automatic Transmission Service Group (ATSG), as well as by highly trained and informed in-house rebuilders. These activities not only ensure that the complex transmission is rebuilt correctly, but funnel the disposal of materials and components to appropriate centers.

Every year, more than two million automatic transmissions are being rebuilt by service organizations. There were 2.1 million rebuilt in 1997 (the last period for complete data), and by 2004, 2.6 million are projected to be rebuilt.[15-1] Although this increase simply reflects the fact that more vehicles are in the marketplace, it also points out the need for reliability improvement.

Significant progress has been made over the years in extending the life of automatic transmission fluid. Where fluid change had been originally recommended at 24,000-mile increments, then 36,000-mile increments and higher, fluid life is now generally accepted, to be 100,000 miles, if not for the life of the vehicle. Actually, fluid life is affected by the severity of vehicle usage. Onboard oxidation indicators have been considered as a possible warning of the need for a transmission fluid change, although they provide an incomplete picture and do not recognize certain other fluid conditions such as contamination.

Transmission fluid is a highly developed, specialized product. It includes additives to resist oxidation and retain lubricating properties and friction characteristics. Fluid operating temperatures are monitored electronically to preclude overheating and fluid degradation. Major transmission overheating can even be signaled by the conventional engine temperature indicator light with appropriate cooling circuit design.

Provisions to contain contamination have also played an important role in extending fluid and transmission life. This is reflected in many designs by the lack of fluid drain plugs in the transmission pan.

Transmission internal components have undergone a similar development, improving capability for environmentally acceptable recycling and durability life. More work definitely needs to be done. Greater application of aluminum and recoverable plastics has replaced cast iron components. Providing more functions-capability to individual components through modern fabricating techniques has tended to reduce the need for many internal seals and gaskets. The seals of special elastomeric materials have achieved cost reductions over metallics in discrete locations but have the disadvantage of needing more special handling during recycling.

References

15-1.   Discussion with Wayne Colonna, Technical Supervisor of the Automatic Transmission Service Group (ATSG), February 26, 2003.

Chapter Sixteen

The Measurables

In the preceding chapters, the components of transmissions were identified, and their advantages and limitations were discussed. How can the best components be culled from the proven storehouse and combined within the boundaries of given vehicle parameters?

The configuration of automotive transmissions has evolved from employing a single manual dry clutch connecting a combination of spur gears on parallel shafts to multiple wet or dry disk clutch members engaging layshaft gearing or multiple planetary gearsets. The torque converter planetary gear transmission has retained its popularity as the number of gear steps has increased. It approaches the ideal performance of an infinitely variable machine, with good reliability. The automatic dual clutch layshaft transmission is gaining acceptance, particularly on the European scene. Other attractive configurations, identified as continuously variable, comprise a variable pulley belt or chain drive, a toroidal drive, or a hydrostatic-type drive, as reviewed in Chapter Four. The hybrid powertrain, comprising an engine and electric motor, and various other power-split configurations offer more fuel economy without sacrificing performance.

In North America, the basic hydrodynamic three-element torque converter transmission retains its leadership, albeit with the addition of a lock-up clutch. Smoothness and durability are the reasons attributed to this longevity, as noted in Chapter Four. Recently published assessments rate the torque converter advantageously as an integral part of most transmission arrangements.

Other transmission components, such as three- and four-member planetary gearing, the multiple-plate clutch, the hydraulic pump, and hydro-electronic controls, continue to be reconfigured but remain as basic building blocks.

All automotive transmissions must conform to industry measurables and meet some common objectives, as discussed next.

1. **Enhance today's highly developed internal combustion engine and help integrate it into a successfully marketed vehicle.**

   In the past, it sometimes seemed that new transmission development went on independently of the engine development. Now, the two are coming together. Powertrain synergism continues to be enhanced with the sharing of engine components, augmented by new developments in electronics and computer control. Thus, vehicle fuel economy gains continue to be made, and engine exhaust emissions continue to be lowered as a result of precise transmission–engine matching and control.

2. **Provide a broad ratio range at high operating efficiency.**

   Transmission engineers have accepted the charge of providing a broader ratio range without sacrificing efficiency. The need for a broad ratio range is fed by the ongoing demand for greater vehicle performance without compromising fuel economy. However, vehicle performance continues to be a function of the vehicle relative power-to-weight ratio. Therefore, the sum of engine power output, attenuated by transmission/drivetrain efficiencies incremented over the entire customer driving life cycle, becomes the true measure of an optimum powertrain for a given engine displacement and vehicle configuration and running weight.

3. **Low parasitic losses.**

   Because conventional passenger vehicles operate mostly at cruising conditions for a significant portion of their life cycle, the transmission design must reflect low parasitic losses because they can have a significant impact on fuel economy, as noted in Chapter Eight. Low parasitic loss is an ongoing target because improvements in modern vehicles to reduce air drag coefficients and rolling friction have continued to lower vehicle road load power requirements, at both city and highway cruising speeds.

   The significance of parasitic loss reduction was demonstrated in Wakamatsu's reporting [16-1] of 1% improvement in vehicle fuel economy due to 20% reduction of transmission drag torque by merely substituting waved clutch plates for flat plates, with related clutch refinements.

4. **Transmission "feel" (not limited to shift feel).**

This objective applies to planetary gear transmissions, dual clutch transmissions, and variator-type transmissions, as well as hybrid vehicle drives. As more electronic power shifts or transfers are added to enhance the ratio range, there is a greater risk of displeasing the driver or passenger by a jerky sensation. Avoiding offending jerkiness through gear ratio step design or integrated computer shift control is only part of the solution. Experience has shown that driver objections to automatic control event changes are more likely if there are "surprise" events such as downshifts occurring as the vehicle enters a grade. This disturbance is more likely with a jump in power going to the wheels, particularly in a vehicle with a numerically low final drive-ratio, or in a light vehicle.

The quality of manual shift engagements from neutral to forward drive or from neutral to reverse is usually not a criterion to which a customer pays any attention when buying a new vehicle. However, these engagements must be smooth and nearly imperceptible without sacrificing a sense of control. A manual "garage shift" into "drive" position is one of the first things a driver does upon starting his car, and continued abrupt engagements can be wearing. These abrupt engagements are even more unacceptable on icy pavement if they result in wheel slip or vehicle skid, especially on an upgrade. To soften such engagements, clutch apply-pressure accumulators are often employed in the design.

The continuously variable transmission (CVT) also is subject to "feel" issues. To ensure smooth startup with a CVT transmission, a torque converter is usually provided to take advantage of its fluid characteristics and its damping characteristics. An alternate viable solution with an earnest following, the substitution of a multi-plate startup clutch carries the baggage of additional intricate controls. These controls must compensate for clutch control sensitivity to temperature—from cold-start engagements to hot engagements such as rocking a vehicle from a stuck condition in mud or a snow bank.

Some experts may downgrade the CVT for its uncoupled startup "feel" or noise. That sensation is often due to controls strategy designed to maintain engine speed close to engine ideal fuel economy–load operating characteristics. The mass/inertia of the belt and pulleys also contribute to that sensation in that type of variator. An algorithm in the engine throttle control module to boost engine torque precisely at a torque demand shift can help offset the inertia effect. Transmission drive control

features such as customer feedback attenuation may be incorporated to help preserve the pleasing drive experience. In most cases, the driver eventually adapts to the "feel" of a new transmission.

5. Reliability.

The competitive vehicle market demands that the transmission provide exemplary service for "the life of the car," a true measurable. Its location in the bowels of the vehicle dissuades the thought of transmission servicing. Confidence in the reliability of a transmission represents a broad process from defining the user's requirements, failure mode and effect (FMEA) analysis, reliability level prediction technique, and qualification testing to feedback, similar to methodology outlined by Popovic.[16-2]

Adherence to the above process prior to design release has yielded improvements in both durability and reliability in recent years. Once in the field, classifying repairs by parameters such as repairs per thousand (R/1000) by miles driven and "things gone wrong" (TGW), and by warranty classification codes (WCC), original equipment manufacturers continue the process by identifying and correcting field service problems and preventing them in the future. Opportunity still abounds. In Chapter Fifteen, we were reminded that more than 2.6 million transmissions are forecasted to be rebuilt in 2004 by independent rebuilders alone.

What are the design parameters necessary to ensure satisfactory reliability life without incurring excessive product cost? To answer this with some confidence requires a projection of the customer duty life cycle for the transmission/vehicle application. That involves summing all the incremental duty-hours of operation at engine full throttle, many more at the various part-throttle conditions, and even more time spent at vehicle road load conditions for each vehicle application, and then accounting for all vehicles to be sold. Of course, each production vehicle and transmission is subjected to different customer usage throughout its lifetime. The range of operation represents the sum of drive engagements, reverse engagements, and full- and part-throttle upshifts and downshifts. This sum depends on the marketing image of the vehicle as well as to whom the vehicle is sold. For example, if a vehicle is built with a numerically low axle ratio or final drive ratio or revolutions per mile per hour (N/V) to meet a fuel economy image, the unperceptive purchaser may be compelled to reach deeper into the throttle more frequently to meet his performance expectations, thus exceeding the design number of transmission downshifts.

The life cycle requirements may be deduced early by obtaining a count of shift occurrences by means of special instrumentation in customer-representative pre-production vehicles. Or it may be discovered too late from field experience at the price of higher warranty cost or eventual lost sales.

Figure 12 illustrates a distributional map of various shift-type occurrences versus vehicle speed for a typical four-speed transmission vehicle test conducted according to a city duty cycle repeated for more than 3,000 miles. It is a powerful graphical technique to visually sort out the distribution of transmission shift events, mostly outside the driver's control, that may adversely impact his driving pleasure. It may also be used for shift durability projection data. In this example, about 27% of the shifts are 2–3 shifts, and 27% are 3–2 downshifts. The 1–2 shifts occurred about 20% of the time, and 2–1 downshifts also about 20%. The less significant shift events are also shown in the figure. Of course, for analytical purposes, this mapping would be supported by a similar distributional map of engine torque at the various shift occurrences.

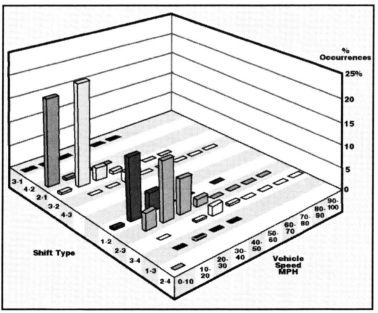

Figure 12. Four-speed transmission shift distribution, city duty cycle.

Graphic: DLB Industrial Design

A newly proposed five- or six-speed transmission could be compared with the preceding four-speed transmission mapping either by computer simulation or by a test run. Shift durability could be analyzed based on this data. The distribution of shifts could be tweaked through control changes.

A similar mapping for a hybrid CVT-type transmission, properly instrumented for transient effective gear ratio, could help pinpoint potential fatigue life issues. A CVT should have durability life exceeding that of the gear shift transmission it seeks to displace. Its driving components such as the push belts are highly stressed, precision assemblies, and a failure can translate into a total loss of vehicle drive.

6. Fuel economy.

It is conceded that a transmission does not burn any fuel, but depending on how the transmission is matched to the engine and vehicle, it can unsuspectingly make or break a vehicle fuel economy image. All the aforementioned transmission objectives of broad ratio range, low parasitic losses, imperceptible shift feel, and durability must be complemented by transmission attributes for good vehicle fuel economy.

Matching the torque converter and transmission gear ratios to the ideal engine fuel island curves at all performance levels has been previously discussed. The CVT transmission must similarly contain the algorithm to control transient ratios at the best fuel economy operation.

The CVT transmission continues to attract more supporters because of its potential for improved engine fuel economy and smoothness. A useful parameter for comparing the CVT to an automatic transmission is the sum of engine revolutions over a standard driving cycle (e.g., the metro-highway). An analysis conducted for a 3.0-liter engine installation showed that a metal belt CVT has an average 3% fewer total revolutions than a five-speed automatic transmission and 11% fewer than a four-speed. This is equivalent to a gain of about 12% in fuel economy.[16-3]

This parameter of total engine revolutions is a simple indicator that, when added to others including acceleration performance, operational flexibility, and power capacity per unit weight, will provide a basis for comparing transmissions.

Customer drive-selection options such as an "economy" drive position featuring lower shift speeds and greater use of the top transmission gears will increase in popularity. The perception that the manual transmission is more economical than an automatic is slowly disappearing with engine calibrations becoming more suitably tailored to the near-constant speed characteristics offered by an automatic.

To study the relative impact of a five-speed transmission, as well as the related powertrain components on vehicle fuel economy, a sort by component as shown in Figure 13 is useful. It illustrates the projected fuel consumption of each vehicle component for the composite metro-highway (M/H) cycle using conventional computer projection analysis based on individual component performance data and physical characteristics data. For this discussion, the value of sorting lies in making a comparison to another transmission; for example, a six-speed transmission (not shown) is projected to yield a greater than 20% improvement in M/H fuel economy with respect to the five-speed.[16-4]

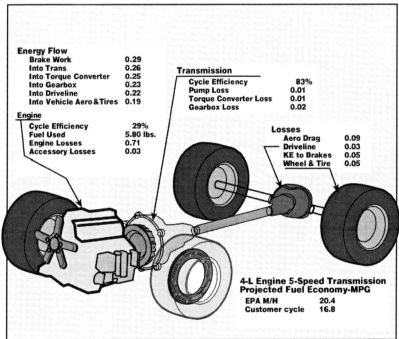

Figure 13. Energy loss by vehicle component (in miles per gallon). (Source: Reference 16-4.)

A transmission is expected to perform flawlessly under a wide range of vehicle operating conditions and ambient temperatures. To prevent overheating, auxiliary coolers are provided in the system. However, the cooling cannot be excessive so as to reduce operating efficiency. Operating temperatures are generally controlled in the range 150–250°F (65–120°C) to maintain acceptable levels of fluid viscosity for transmission efficiency and component function and not degrade the fluid when operating at high temperatures. Experience has shown that with an increase of 17°F in average transmission fluid temperature can double the rate of fluid deterioration, as measured by laboratory oxidation tests.[16-5] Translating that estimate to a 30,000-mile transmission operating at 275°F, lowering the operating oil temperature to 250°F would extend fluid life to about 75,000 miles. However, there are many other measurable fluid properties to be controlled that affect durability life, such as resistance to sludging, retention of friction characteristics, resistance to viscosity degradation, and accommodation to contamination.

Thermistor electronic controls with appropriate circuitry help heat the hydrodynamic-type transmission up to normal operating temperature faster from cold starts. At the other extreme, to prevent overheating with heavy usage requires a large fluid heat exchanger. The CVT transmission has a slightly different temperature control problem: dissipating heat buildup from the high-pressure energy that is required to rapidly change the pulley ratio. In those installations, new non-hydrocarbon friction fluids are showing promise in being more tolerant to temperature extremes.

References

16-1.   Wakamatsu, H., et al., "Honda's 5-Speed All Clutch to Clutch Automatic Transmission," SAE Paper No. 2002-01-0932, Society of Automotive Engineers, Warrendale, PA, 2002.

16-2.   Popovic, P., and Ivanovic, G., "The Methodology for Reliability Design of Power Transmission," SAE Paper No. 2002-01-2201, Society of Automotive Engineers, Warrendale, PA, 2002.

16-3.   Aboo, K., and Kobayashi, M., "Development of a New Metal Belt CVT for High Torque Engines," SAE Paper No. 2000-01-0829, Society of Automotive Engineers, Warrendale, PA, 2000.

16-4. Interview with R.C. Roethler, Senior Technical Specialist, Ford Motor Company, May 22, 2003.

16-5. Griffin, T.J., "Temperature Effect on Transmission Operation," in *Design Practices: Passenger Car Automatic Transmissions,* AE-5, Society of Automotive Engineers, Warrendale, PA, 1962.

Chapter Seventeen

The "New" Transmission

How do we apply all these measurables to come up with the most value-effective transmission? The answer somewhat depends on the vehicle application, such as passenger car, pickup truck, SUV, and so forth, as discussed in Chapter Two. However, some general direction may be drawn.

First, addressing the layshaft gear mechanism of tomorrow, the manual transmission in four- to six-speed configurations will continue to be the gearing of choice by the sporty, demanding driver of either the passenger car or light truck because of the implied fuel economy and durability. The feel of responsiveness and control will have its reward. It will take continued attention to the optimum design of the clutch and damper assembly as well as synchronizer smoothness to meet the expectations of this "manual customer."

A step up from the basic layshaft manual, the automated manual transmission (AMT) as discussed in Chapter Four will be the transmission of choice for its performance with no torque interruption for the enthusiast. He will be pleasantly surprised by the good fuel economy as well. Exemplified in the 2003 Volkswagen six-speed direct shifting gearbox (DSG), this AMT popularizes the dual wet clutch configuration (DualTronic by Borg Warner). Smooth six-speed coverage, assured with advances in electronic clutch control and a well-tuned spring damper element, eliminates the need for a torque converter in that application. Projected volume is 20% of total European production by the year 2010.[17-1]

The distinction between the AMT-type six-speed with layshaft gearing and a six-speed planetary gear unit can be blurred at first glance. However, the perceptive customer will recognize some significant differences.

For example, will the workhorse for the all-around passenger vehicle continue to be the multiple-gear planetary transmission? The planetary gear arrangement, with its gears in constant mesh, obviates the need for the usual four or more precision synchronizers. Also, the planetary configuration, with lock-up torque converter, approaches infinitely variable performance.

Figure 14 illustrates a vehicle full-throttle tractive effort curve projected with the same basic five-speed torque converter planetary gear transmission and the same engine used in Figure 11 of Chapter Twelve to illustrate power loss. An infinitely variable transmission (IVT) tractive effort curve is shown for comparison purposes. It is the uppermost smooth heavy curve.

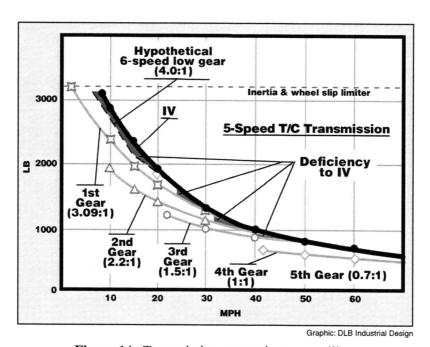

Figure 14. Transmission gear ratio range ceiling.

The shaded areas at the 1–2, 2–3, and 3–4 shift areas graphically illustrate the performance deficiencies of the five-speed transmission having 3.09:1 first-gear-ratio with respect to an IVT.

To represent the level of full-throttle performance that would normally be available with a six-speed planetary transmission, a curve of 4:1 first gear ratio was included in Figure 14. It can be seen that the full-throttle performance of a six-speed planetary transmission closely approaches performance of an IVT. This is particularly significant because six is a practical limit to the number of planetary gear steps. As described in Chapter Five, increasing the number of planetary transmission gear steps beyond six—to seven, eight, or more (except eleven)—results in component design redundancy and possibly unacceptably high component relative speeds at full-throttle shift points.

Therefore, the six-speed torque converter planetary transmission should be around for a long time until it is displaced by completely unique powertrain architecture.

Figure 15 is a schematic of the original configuration of the series-parallel hybrid drive such as found in the Toyota Prius front-wheel-drive vehicle available in North America since the year 2000.

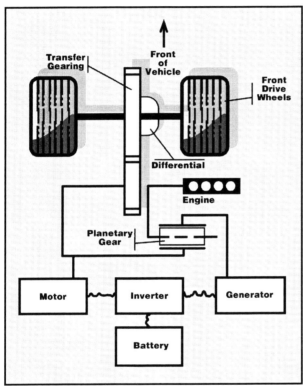

Graphic: DLB Industrial Design

Figure 15. Series/parallel hybrid system. (Source: SAE TOPTEC, August 12–13, 2003.)

New developments improving this drive with uniquely controlled low- and high-voltage motors/inverters, along with engine power upgrade, have been reported.[17-2] The resulting upgrade in overall power and efficiency is reportedly sufficient for adaptation of this hybrid configuration to the mid-size vehicle. This system indicates potential to become a viable alternative to the conventional internal combustion engine–transmission drive.

Of course, the direction that future automotive transmission configurations will take is closely related to the design of the powerplant and the vehicle, both subject to global energy supply conditions. Beyond the baggage of existing capital investment,

future automotive transmissions will continue to be dependent on the economics of delivering fuels to power the engines for both passenger and freight transportation needs. The conventional fuels of today, gasoline and diesel, have viable alternatives such as electricity, hydrogen, compressed natural gas, and methanol, to name only a few. These alternatives were analyzed by Kreith and West [17-3] on a "well-to-wheel" basis. They projected that the highest overall process efficiency, 32% well-to-wheel, exists from hybrid diesel–natural gas power and also from a hybrid spark ignition engine option. Various fuel cell drive arrangements and electric-battery combinations were ranked stages lower after taking into account the manufacturing process cycle needed to produce and deliver this energy. The fuel cell hydrogen system had the lowest ranked efficiency, which was 13%. They noted that 13% is about equivalent to one of today's SUVs. Of course, there are various possible energy system configurations that represent efficiencies between 13% and 32%, depending on the combination of powertrain and global energy supply.

What does all this mean to the designer of tomorrow's transmission if "well-to-wheel" efficiency were to become the predominate consideration? Based on published economy improvements that are taking place relative to current powertrains, hybrid power-split configurations could be at the top of the list. The transmission would complement the somewhat smaller engine, in hybrid architecture, matched with a motor generator drive. The gear ratio span, including the transfer gear, could be somewhat lower because of the broader operating torque range of the latest electronically controlled engines. Although ancillary features such as "park" will likely continue as a transmission function, "hill brake" could be assumed by the motor generator. The shift selector/console will be further enhanced with various performance and economy-range customer features. Obviating the torque converter, the versatile electric motor will provide smooth startup as well as reverse drive capability.

Extrapolating current design trends indicates that mass-production passenger car and truck configurations will be headed down the following roads:

- Engine located predominately in the front of the vehicle.

- Expanded production of six-speed torque converter planetary transmission applications, both in front- and rear-wheel-drive vehicles.

- Growth of multi-clutch automated layshaft transmission applications, particularly in Europe.

- Continuously variable transmissions (CVT) with push belt-type variators strengthening their hold in light-vehicle applications due to their smooth operation and fuel economy advantage.

- Continued production of existing four-speed planetary transmission applications with ongoing cost reduction improvements to amortize existing investment.

- Increasing production of power-split hybrid vehicles using next-generation internal combustion engines/planetary transmissions in parallel with flexible voltage motor generator and more efficient electric power regeneration to reduce battery capacity requirements.

- Application of electromagnetic starter motors, taking advantage of the potential for energy capture.

- Increasing numbers of models of crossover vehicles with four-wheel drive (4WD) on demand.

- Increased interest in a hybrid truck 4WD arrangement with an additional electronically controlled electric motor driving the rear wheels. There may be opportunity for an electric motor at each rear wheel, integrating differential rear-wheel speeds with the steering function to facilitate "park" and "backup" maneuvers.

- Continued development of the hydrogen fuel cell powertrain and energy transfer.

- Continued interest in split torque configurations of engine–hydrostatic variators in trucks to take advantage of flexibility of location within the vehicle.

Developments will accelerate to reconcile the extremes in vehicle utilization between city start-and-stop driving and suburban-type driving. In start-and-stop driving, as in city delivery vehicles, the features of hybrid parallel-type drives are of particular advantage because of the conservation of coast and braking energy.

Improvements to the CVT-type of transmission will continue. Issues such as slower response to demand-shifts and weight should be resolvable with lighter, stronger materials and more refined controls logic. For example, the capability to rock the vehicle out of a snow bank is compromised by the inertia effect in friction drive CVTs. Offsetting techniques, such as engine throttle boosts or split torque de-clutching, should be viable solutions.

In summary, significant opportunities are available to design the "new" transmission. These must dovetail with the objective of improving overall vehicle power-to-weight to meet performance without sacrificing fuel economy or clean air requirements.

Ongoing year-to-year improvement actions probably will not be enough to meet the demands of society for the long range. Examples that come to mind are reducing transmission package size and the corresponding weight needed to transmit the torque. Lighter, more exotic materials such as magnesium and carbon fiber and polymers will help to meet the demand for more vehicle performance and more space-efficient vehicle packaging.

Not enough!

New paradigms are facilitating significant engine–transmission–vehicle improvements. The automated manual transmission (AMT) has injected new life into the aging stick-shift. Hybrid power-split technology has raised the bar of operating efficiency. Rapid developments in hydrogen fuel cell technology, spearheaded by military and space programs, are advancing the fuel cell vehicle closer to mass production. Projections of return on investment (ROI) and efficiencies continue to improve, in spite of less than optimistic well-to-wheel energy numbers.

Future transmissions and powertrain design trends will be impacted by many factors, as has been stated. Figure 16 is an attempt to look into the next decade or two, and it illustrates anticipated increase in market share of hybrid gasoline and diesel installations continuing with a simplified form of the current multi-gear automatic transmission. The CVT will continue to make inroads at the expense of the current torque converter transmission. Fuel cell and solar applications can be expected to debut by the year 2020, bringing their own design and infrastructure issues.

Actually, today's consumer is relatively satisfied with the performance of his vehicle. He is hardly cognizant of the transmission in his vehicle. Before major powertrain changes can become completely accepted, there is much opportunity for further innovation so that current overall vehicle performance, durability, and "fun-to-drive" characteristics are not compromised.

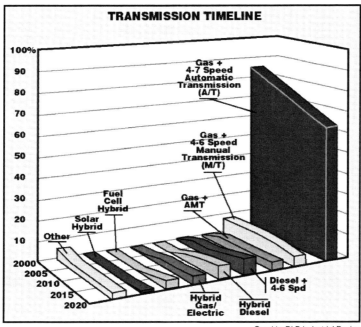

Figure 16. Passenger car transmission time line.

Graphic: DLB Industrial Design

References

17-1.   Nitsche, Joerg, "Dual Clutch Transmission: A Fun-to Drive and Fuel Efficient Automatic Transmission," SAE TOPTEC, August 12–13, 2003, Society of Automotive Engineers, Warrendale, PA, 2003.

17-2.   Hermance, D., "Toyota HEV Transmission," SAE TOPTEC, August 12–13, 2003, Society of Automotive Engineers, Warrendale, PA, 2003.

17-3.   Kreith, F., and West, R.E., "Gauging Efficiency Well-to-Wheel," *Mechanical Engineering Power 2003*, pp. 20–23.

Appendix One

Significant Transmission Innovations

The following is an attempt to list *significant* transmission innovations that appeared in mass production to my knowledge. Such a list is always subject to updating.

Date of Introduction	Innovation	Application
1908	Two-speed planetary gearset	Model T transmission
1928	Synchronized shift manual transmission	Cadillac Synchromesh
1930	Integral free-wheeling manual transmission	Studebaker
1933	Self-shifting two-speed transmission	Reo Self Shifter
	Vacuum-operated clutch manual transmission	Chrysler
1934	Overdrive in three-speed manual	Buick
1938	Pre-selective electric gear shift	Packard, Hudson
1939	Fluid coupling with three-speed manual gear box	Chrysler Fluid Drive
1940	Four-speed planetary 62% mechanical split torque A/T with fluid coupling	General Motors Hydramatic
1946	Semi-automatic countershaft transmission	Chrysler Hydrodrive
1948	Five-element torque converter (T/C) two-speed transmission. Pawl park brake.	Buick Dynaflow
1949	Lock-up clutch T/C two-speed transmission	Packard Ultramatic
	Three-element welded T/C with two simple gearsets, non-synchronous shifts	Studebaker DG200
1950	Three-speed Ravigneau gearset transmission with air-cooled three-element T/C	Ford Ford-O-Matic
	Four-speed countershaft transmission with three-element T/C	Chrysler M-6

Date of Introduction	Innovation	Application
1952	Variable capacity fluid pump	General Motors Hydramatic
1954	Four-element welded T/C	Chrysler Powerflight
	Twin-turbine variable pitch stator T/C transmission	Buick Dynaflow
1956	Second fluid coupling, used as dump and fill clutch with four-speed planetary gearset	General Motors Stratoflight
1957	Five-element T/C with three-stage turbine no-shift transmission	Chevrolet Turbo-Glide
	Three-speed transmission with Simpson gearset	Chrysler TorqueFlite
1958	"Shift by wire," pushbutton shift on steering wheel	Edsel Teletouch
1960	Two-speed A/T, rear-wheel-drive transaxle	Chevrolet Corvair
1961	Two-speed planetary transmission with stator reverse	Buick Dual-Pitch
1965	Dual-axis longitudinal front-wheel-drive A/T with "HyVo" chain	Hydramatic 425
1966	Non-synchronous 1–2 shift, three-speed A/T	Lincoln C6
1972	Three-speed all-clutch transmission (no bands)	Aisin AW 355
1978	Transverse three-parallel axis transmission, with geared drive	Chrysler A404
1980	Integral overdrive four-speed A/T, 60% mechanical split torque in third gear	Ford AOD
	Transverse parallel-axis chain drive front-wheel-drive A/T	General Motors Hydramatic 125
1981	93/7% split torque in top gear, twin parallel axis front-wheel-drive transmission	Ford ATX "wide ratio"
1983	Four-speed A/T employing only two simple gearsets	General Motors 700-R4
1984	Centrifugal clutch torque converter with three-speed A/T	Ford C5
	U-drive transmission configuration	General Motors TH440
1989	Feedback controls	Chrysler A604
1990	Integrated transmission/engine shift control (five-speed)	Nissan JATCO

Date of Introduction	Innovation	Application
1999	CVT traction drive	Nissan JATCO
2001	Wet clutch drive input to metal belt CVT	Ford Transmatic
	Electromagnetic drive input to metal belt CVT	Subaru ECVT
2002	Change-by-wire six-speed clutch transmission	Mercedes Sequentronic
	Six-speed longitudinal fully synchronized A/T with one simple and one Ravigneau gearset with only five friction members	ZF
2003	Dual clutch six-speed layshaft transmission (DCT)	Volkswagen Golf R32

List of Acronyms

2WD	Two-wheel drive
4WD	Four-wheel drive
AMT	Automated manual transmission
ASM	Automated shift mechanism
AST	Automated shift transmission
A/T	Automatic transmission
ATF	Automatic transmission fluid
ATSG	Automatic Transmission Service Group
AWD	All-wheel drive
BSCF	Brake specific fuel consumption
CPU	Central processing unit
CV	Constant velocity (as in CV joints)
CVT	Continuously variable transmission
DCT	Dual clutch transmission
DSG	Direct shifting gearbox
ECU	Electronic control unit
FEM	Finite element method
FMEA	Failure mode and effect analysis

FWD	Front-wheel drive
IVT	Infinitely variable transmission
N/V	Number (driveshaft) of revolutions to velocity (mph)
NVH	Noise, vibration, and harshness
OWC	One-way clutch
PWM	Pulse-width modulated (solenoid)
ROI	Return on investment
RWD	Rear-wheel drive
SUV	Sport utility vehicle
T/C	Torque converter
TGW	"Things gone wrong"
WCC	Warranty classification code

About the Author

Martin G. Gabriel, or Marty as he is known, has demonstrated a flair for writing since his college days when he was feature editor of the collegiate newspaper, *Technology News*, while at the Illinois Institute of Technology (IIT) in Chicago. He received his B.S.M.E. degree there in 1947 and later went on to receive his M.S.E.M. degree at the University of Michigan in 1955.

Beginning his working career at Borg Warner Corporation in the Chicago area, Marty's interest in torque converter research was whetted there under the direction of V.J. Jandasek. Marty was fortunate to have had the opportunity to study under the tutelage of Ernst W. Spannhake of Trilok Research Society of Germany, where Spannhake refined the Schneider torque converter system, and emigrated after World War II to briefly teach at IIT. While at Borg Warner, Marty contributed to the intense development of what was to become the first Ford production automatic transmission, the Ford-O-Matic.

In 1950, Marty joined Jandasek at Ford Research in Dearborn, and after a varied career in torque converters and transmissions and reliability engineering, he retired from Ford in 1995 as senior reliability engineer. He is the author of twelve U.S. patents in the field, including a variable-speed multi-element converter that obviated the need for any gears. His later Ford assignments included spearheading the first powertrain activity to support 100,000-mile objectives and reliability programs in the powertrain quality office.

Marty's long association with SAE International began in 1950. He was president of the National Junior Activity, and he continues his active work on various SAE committees to this day. Past chairman of the EMB and SAE Transmission Committees, he contributed to the development of many standards, as well as to the SAE publication of *Design Practices: Passenger Car Automatic Transmissions* (AE-5), which was first published in 1962. Under his chairmanship of the SAE Forum Committee, that design practice was completely updated and published as AE-18 in 1988. A past elected member of the SAE board of directors, Marty continues to serve on the SAE Continuing Professional Development Group responsible for meeting seminars and TOPTECS.

Marty is a registered Professional Engineer of Michigan since 1951 and an active member of several professional and civic organizations. He taught "Automotive Dynamics" during the evening at Oakland University. He has received many awards, including "Engineer of the Year" in 1994 and the Forest McFarland Award. He is past president of the Michigan Society of Professional Engineers, a member of the Florida Society of Professional Engineers, and a life member of ASME International and SAE International. He continues to serve on the boards of the Detroit Area Council of the Boy Scouts of America (BSA), and the Catholic Youth Organization (CYO) serving more than 100,000 youth and volunteers of the Greater Detroit Area, of which he is past president. He is an active member of the Parish Council of his church, St. Owen.

Marty's hobbies include photography, a backyard fruit orchard, and portrait painting, for which he has received honorable mention in juried exhibits. He and his wife, Marie, have been married for 54 years and have five adult children and seven grandchildren.